STERLING
Test Prep

ACS Organic Chemistry

Practice Questions

4th edition

4   3   2   1

ISBN-13: 979-8-8855710-9-8

Sterling Test Prep materials are available at quantity discounts.
Contact info@sterling–prep.com

Sterling Test Prep
6 Liberty Square #11
Boston, MA 02109

© 2023 Sterling Test Prep

Published by Sterling Test Prep

 Printed in the U.S.A.

STERLING
Test Prep

Thousands of students use our study aids to achieve high test scores!

Scoring well on ACS Organic Chemistry exam is a challenging task. This book helps you develop and apply knowledge to quickly choose the correct answer on the test. Solving targeted practice questions builds your understanding of fundamental general chemistry concepts and is a more effective strategy than merely memorizing terms.

This book provides high-yield practice questions covering organic chemistry topics. Chemistry instructors with years of teaching experience prepared these questions by analyzing the test content and developing practice material that builds your knowledge and skills crucial for success on the ACS. Our test preparation experts structured the content to match the current test requirements.

The detailed explanations describe why an answer is correct and – more important for your learning – why another attractive choice is wrong. They provide step-by-step solutions and teach the important details of organic chemistry mechanisms and reactions needed to answer ACS exam questions. Read the explanations carefully to understand how they apply to the question and learn important organic chemistry principles and the relationships between them.

With this practice material, you will significantly improve your test score.

230105frd

ACS General Chemistry Review provides comprehensive and targeted coverage of general chemistry topics tested on ACS. The content covers foundational principles and theories necessary to understand the material and answer test questions.

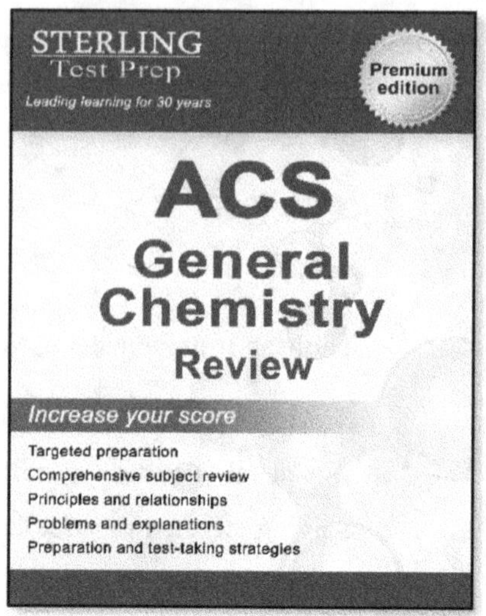

ACS General Chemistry Practice Questions provides high-yield practice questions covering topics tested on ACS General Chemistry exam. Develop the ability to apply your knowledge and quickly choose the correct answer to increase your test score.

Visit our Amazon store

Table of Contents

Tested topics: nomenclature, covalent bonds, stereochemistry, molecular structure and spectra, separation and purification, alkanes & alkyl halides, alkenes, alkynes, aromatic compounds, alcohols, aldehydes & ketones, carboxylic acids, carboxylic acid derivatives, amines.

College Level Examination Program (CLEP)

Biology Review

Biology Practice Questions

Introductory Psychology Review

American Government Review

College Algebra Practice Questions

College Mathematics Practice Questions

History of the United States I Review

History of the United States II Review

Western Civilization I Review

Western Civilization II Review

Social Sciences and History Review

Introductory Business Law Review

Visit our Amazon store

ACS Organic Chemistry Preparation and Test-Taking Strategies

Test preparation strategies

The best way to do well in ACS Organic Chemistry is to be good at organic chemistry. There is no way around knowing the subject; proper preparation is key to success. Prepare to answer with confidence as many questions as possible.

Study in advance

Assuming you completed all necessary classwork and did fairly well at it, devote 2 to 3 months to studying. The information is manageable by studying regularly before the test.

Cramming is not a successful tactic. However, do not study too far in advance. Studying more than four months ahead is not advised and may result in fatigue and poor knowledge retention.

Develop a realistic study and practice schedule

Cramming eight hours a day is unfeasible and leads to burnout, which is detrimental to performance.

Commit to a realistic study and practice schedule.

Remove distractions

During this preparation period, temporarily eliminate distractions.

However, balance is critical, and it is crucial not to neglect physical well-being and social or family life.

Prepare with full intensity but do not jeopardize your health or emotional well-being.

Develop an understanding over memorization

When studying, devote time to each topic.

After a study session, write a short concept outline to clarify relationships and increase knowledge retention.

Make flashcards

Avoid commercial flashcards because making cards help build and retain knowledge.

Consider using self-made flashcards to develop knowledge retention and quiz what you know.

Find a study partner

Occasionally studying with a friend to prepare for the test can motivate and provide accountability.

Explaining concepts to another improves and fine-tunes your understanding, integrates knowledge, bolsters competence, and identifies deficiencies in comprehension.

Take practice tests

Do not take practice tests too early.

First, develop a comprehensive understanding of concepts.

In the last weeks, use practice tests to fine-tune your final preparation. If you are not scoring well on practice tests, you want time to improve without undue stress.

Alternate studying and practicing to increase knowledge retention and identify areas for study.

Taking practice tests accustoms you to the challenges of test-taking.

Test day strategies

Be well-rested and eat the right foods

Get a full night's sleep before the test for proper mental and physical capacity. If you are up late the night before, you will have difficulty concentrating and focusing on the test.

Avoid foods and drinks that lead to drowsiness (carbohydrates and protein).

Avoid drinks high in sugar, causing glucose to spike and crash.

Pack in advance

Check what you are allowed to bring to the test. Pay attention to the required check-in items (e.g., confirmation, and identification).

Pack the day before to avoid the stress of not frantically looking for things on test day.

Arrive at the testing center early

Starting right is an advantage. Allow time to check in and remain calm before the test begins.

Plan your route to the center correctly without additional challenges and unnecessary stress. Map and test the route to the center in advance and determine needed parking.

If you are unfamiliar with the test location, visit before the test day to practice and avoid travel errors.

Maintain a positive attitude

Avoid falling into a mental spiral of negative emotions. Too much worry leads to underperformance.

If you become anxious, chances are higher for lower performance in preparation and during the test.

Inner peace helps during preparation and the high-stakes test.

To do well on the test requires logical, systematic, and analytical thinking, so relax and remain calm.

Focus on progress

Do not be concerned with other test-takers.

Someone proceeding rapidly through the exam may be rushing or guessing on questions.

Take breaks

Do not skip the available timed breaks. Refreshing breaks help you finish strong.

Eat a light snack to replenish your energy. Your mind and body will appreciate the available breaks.

The best approach to any test is *not* to keep your head down the whole session.

While there is no time to waste, breathe deeply for a few seconds between questions.

Momentarily clear your thoughts to relax your mind and muscles.

Time management strategies

Timing

Besides good preparation, time management is the critical strategy for the exam.

Timed practice is not the objective at this stage.

While practicing, note how many questions you would have completed in the allotted time.

Average time per question

In advance, determine the average time allotted for each question.

Use two different approaches depending on which preparation phase you are working.

During the first preparation phase, acquire, fortify, and refine your knowledge.

During final practice, use the time needed to develop analytical thought processes related to specific questions.

Work systematically

Note your comprehension compared to the correct answers to learn and identify conceptual weaknesses.

Do not overlook explanations to questions as a source of content, analysis, and interdependent relationships.

During the second preparation phase, do not spend more than the average allotted time on each question when taking practice tests.

Pace your response time to develop a consistent pace and complete the test within the allotted time.

If you are time-constrained during the final practice phase, work more efficiently, or your score will suffer.

Focus on the easy questions and skip the unfamiliar

Easy or difficult questions are worth the same points. Score more points for three quickly answered questions than one hard-earned victory.

Answer the familiar and easy questions to maximize points if time runs out.

Identify strengths and weaknesses

Skip unfamiliar questions in the first round, as challenging questions require more than average time.

In the second review, questions you cannot approach systematically or lack fundamental knowledge will not be answered through analysis.

Use the elimination and educated guessing strategy to select an answer and move on to another question.

Do not overinvest in any question

Some questions consume more time than the average and make you think of investing more time.

Stop thinking that investing more time in challenging questions is productive.

Do not get entangled with questions while losing track of time.

The test is timed, so do not spend too much time on any questions.

Look at every question on the exam

It is unfortunate not to earn points for a question you could have quickly answered because you did not see it.

If you are in the first half of the test and spending more than the average on a question, select the best option, note the question number, and move on.

Do not rush through the remaining questions, causing you to miss more answers.

If time allows, return to marked questions and take a fresh look.

However, do not change the original answer unless you have a reason to change it.

University studies on students show that hastily changing answers often replace correct and incorrect answers.

Multiple-choice questions

Answer all questions

How many questions are correct, not how much work went into selecting the answers matters. An educated guess earns the same points as an answer known with confidence.

On the test, you need to think and analyze information quickly.

The skill of analyzing information quickly cannot be gained from a college course, prep course, or textbook.

Working efficiently and effectively is a skill developed through focused effort and applied practice.

Strategic approach

Strategies, approaches, and perspectives for answering multiple-choice questions help maximize points.

Many strategies seem like common sense. However, these helpful approaches might be overlooked under the pressure of a timed test.

While no strategy replaces comprehensive preparation, apply probability for success on unfamiliar questions.

Understand the question

Know what the question is asking before selecting an answer.

It is surprising when students do not read (and reread) carefully and rush to select the wrong answer.

The test-makers anticipate hasty mistakes, and many enticing answers include specious choices.

A successful student reads and understands the question precisely before looking at the answers.

Focus on the answer

Separate the vital information from distracters and understand the design and thrust of the question.

Answer the question and not merely pick a factually accurate statement or answer a misconstrued question.

Rephrasing the question helps articulate precisely what the correct response requires.

When rephrasing, do not change the question's meaning; assume it is direct and to the point as written.

After selecting the answer, review the question and verify that the choice selected answers the question.

Answer the question before looking at choices

This valuable strategy is applicable if the question asks for generalized factual details.

Form a thought response first, then look for the choice that matches your preordained answer.

Select the predetermined statement as it is likely correct.

Factually correct, but wrong

Questions often have incorrect choices that are factually correct but do not answer the question.

Predetermine the answer and do not choose merely a factually correct statement.

Verify that the choice answers the question.

Do not fall for the familiar

When in doubt, it is comforting to choose what is familiar. If you recognize a term or concept, you may be tempted to pick that choice impetuously.

However, do not go with familiar answers merely because they are familiar.

Before selecting the answer and how it relates to the question, think about it.

Experiments questions

Determine the purpose, methods, variables, and controls of the experiment.

Understanding the information presented helps answer the question.

With multiple experiments, understand variations of the same experiment by focusing on the differences.

For example, focus on changes between the first and second experiments, second and third, and first and third.

Direct comparison between experiments helps organize and apply the information to the answer.

Words of caution

The words *"all," "none,"* and *"except"* require attention. Be alert with questions containing these words, as the answer may not be apparent on the first read of the question.

Double-check the question

After selecting an answer, return to the question to ensure the selected choice answers the question as asked.

Fill the answers carefully

Many mistakes happen when filling in answers.

Filling answers correctly is simple but crucial.

Note the question number and enter answers accordingly. If you skip a question, skip it on the answer sheet.

Quantitative multiple-choice questions

Know the equations

Many exam questions require scientific equations, so understand when to use each.

As you work with this book, learn to apply formulas and equations, and use them in many questions.

Manipulate the formulas

Know how to rearrange the formulas.

Many questions require manipulating equations to calculate the correct answer.

Familiarity includes manipulating the terms, understanding relationships, and isolating variables.

Estimating

Estimating helps choose the answer quickly for quantitative questions with a sense of the order of magnitude.

Often, estimation enables the correct answer to be identified quickly compared to the time needed for calculations.

Estimating is especially applicable to questions where the answer choices have different orders of magnitude.

It saves time to estimate instead of computing the solution.

Evaluate the units

For quantitative problems, analyze the units to build relationships between the question and the correct answer.

Understand what value is sought and eliminate wrong choices with improper units.

Make visual notes

Write, draw, or graph the question to determine the information provided, the objective, and the concept tested.

A chart or table often makes the solution apparent, even if a question does not require a graphic answer.

Elimination strategies

If the correct answer is not immediately apparent, use the process of elimination.

Use the strategy of educated guessing by eliminating one or two answers.

Usually, at least one answer choice is easily identified as wrong. Eliminating one choice increases the odds of selecting the correct one.

Process of elimination

Eliminate choices:

- Use proportionality for quantitative questions to eliminate choices too high or too low.

- Eliminate answers that are *almost right* or *half right*.

  Consider *half right* as *wrong* since distractor choices are purposely included.

- If two answers are direct opposites, the correct answer is likely one of them.

  However, note if they are direct opposites or another reason to consider them correct. Therefore, eliminate the other choices and narrow the search for the correct one.

- For numerical questions, eliminate the smallest and largest numbers (unless for a reason).

Roman numeral questions

Roman numeral questions present several statements and ask which is/are correct.

These questions are tricky for most test-takers because they have more than one potentially correct statement.

Roman numeral questions are often included in combinations with more than one answer.

Eliminating a wrong Roman numeral statement eliminates choices that include it.

Educated guessing

Correct ways to guess

Do not assume you must get every question right; this will add unnecessary stress during the exam.

You will (likely) need to guess for some questions.

Answer as many questions correctly as possible without wasting time.

For challenging questions, random guessing does not help. Use educated guessing after elimination.

Playing the odds

Guessing is a form of "partial credit" because while you might not be sure of the correct answer, you have the relevant knowledge to identify some wrong choices.

There is a 25% chance of correctly guessing random responses since questions have four choices. Therefore, the odds are guessing 1 question correctly to 3 incorrectly.

Guessing after elimination of answers

After eliminating one answer as wrong, you have a 33% chance of a lucky guess. Therefore, your odds move from 1 question right to 2 questions wrong.

While this may not seem like a dramatic increase, it can make an appreciable difference in your score.

Confidently eliminating two wrong choices increases the chances of guessing correctly to 50%!

When using elimination:

- Do not rely on gut feelings alone to answer questions quickly.

  Understand and recognize the difference between *knowing* and *gut feeling* about the answer.

  Gut feelings should sparingly be used after the process of elimination.

- Do not fall for answers that sound "clever," and choose "bizarre" answers.

  Choose them only with a reason to believe they may be correct.

Eliminating Roman numeral choices

A workable strategy for Roman numeral questions is to guess the wrong statement.

For example: A. I only

B. III only

C. I and II only

D. I and III only

Notice that statement II does not have an answer dedicated to it. This indicates that statement II is likely wrong and eliminates choice C, narrowing your search to three choices.

However, if you are confident that statement II is the answer, do not apply this strategy.

Diagnostic Tests

Diagnostic Test 1

This Diagnostic Test is designed to assess your proficiency on each topic and NOT to mimic the test. Use your test results and identify areas of strength and weakness to adjust your study plan and enhance your knowledge.

#	Answer:				Review	#	Answer:				Review
1:	A	B	C	D	___	36:	A	B	C	D	___
2:	A	B	C	D	___	37:	A	B	C	D	___
3:	A	B	C	D	___	38:	A	B	C	D	___
4:	A	B	C	D	___	39:	A	B	C	D	___
5:	A	B	C	D	___	40:	A	B	C	D	___
6:	A	B	C	D	___	41:	A	B	C	D	___
7:	A	B	C	D	___	42:	A	B	C	D	___
8:	A	B	C	D	___	43:	A	B	C	D	___
9:	A	B	C	D	___	44:	A	B	C	D	___
10:	A	B	C	D	___	45:	A	B	C	D	___
11:	A	B	C	D	___	46:	A	B	C	D	___
12:	A	B	C	D	___	47:	A	B	C	D	___
13:	A	B	C	D	___	48:	A	B	C	D	___
14:	A	B	C	D	___	49:	A	B	C	D	___
15:	A	B	C	D	___	50:	A	B	C	D	___
16:	A	B	C	D	___	51:	A	B	C	D	___
17:	A	B	C	D	___	52:	A	B	C	D	___
18:	A	B	C	D	___	53:	A	B	C	D	___
19:	A	B	C	D	___	54:	A	B	C	D	___
20:	A	B	C	D	___	55:	A	B	C	D	___
21:	A	B	C	D	___	56:	A	B	C	D	___
22:	A	B	C	D	___	57:	A	B	C	D	___
23:	A	B	C	D	___	58:	A	B	C	D	___
24:	A	B	C	D	___	59:	A	B	C	D	___
25:	A	B	C	D	___	60:	A	B	C	D	___
26:	A	B	C	D	___	61:	A	B	C	D	___
27:	A	B	C	D	___	62:	A	B	C	D	___
28:	A	B	C	D	___	63:	A	B	C	D	___
29:	A	B	C	D	___	64:	A	B	C	D	___
30:	A	B	C	D	___	65:	A	B	C	D	___
31:	A	B	C	D	___	66:	A	B	C	D	___
32:	A	B	C	D	___	67:	A	B	C	D	___
33:	A	B	C	D	___	68:	A	B	C	D	___
34:	A	B	C	D	___	69:	A	B	C	D	___
35:	A	B	C	D	___	70:	A	B	C	D	___

This page is intentionally left blank

1. Which of the following is the IUPAC name for this compound?

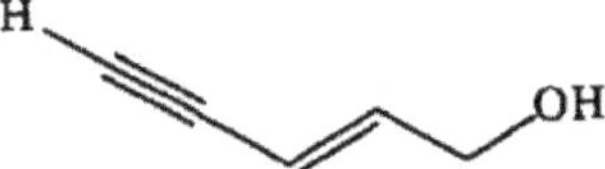

A. (*Z*)-pent-3-en-1-yn-5-ol

B. (*E*)-pent-3-en-1-yn-5-ol

C. (*Z*)-pent-2-en-4-yn-1-ol

D. (*E*)-pent-2-en-4-yn-1-ol

2. Identify the number of carbon atoms for each hybridization in the molecule.

$O=CH–CH_2–CH=C=C=CH_2$

	sp	*sp*^{2}	*sp*^{3}
A.	1	4	1
B.	2	3	1
C.	0	3	3
D.	1	3	2

3. What is the relationship between the following molecules?

A. Diastereomers

B. Structural isomers

C. Identical

D. Enantiomers

4. Which of the following is NOT true?

A. NMR spectroscopy utilizes magnetic fields

B. IR spectroscopy utilizes light of wavelength 200-400 nm

C. UV spectroscopy utilizes molecular vibrations within molecules

D. Mass spectrometry utilizes fragmentation of the sample

5. In thin-layer chromatography (TLC), which of the following best describes the behavior of a polar component within a sample in an ether solvent?

    A. The polar component moves downward compared to the solvent

    B. The polar component moves a similar distance compared to the solvent

    C. The polar component moves less distance compared to the solvent

    D. Neither the solvent nor the polar component moves because of the differences in polarity

6. Which of the following undergoes bimolecular nucleophilic substitution at the fastest rate?

    A. 1-chloro-2,2-diethylcyclopentane   **C.** 1-chlorocyclopentene

    B. 1-chlorocyclopentane   **D.** 1-chloro-1-ethylcyclopentane

7. What is the major product formed from the reaction of 2-bromo-2-methylpentane with sodium ethoxide?

    A. 2-methylpent-2-ene   **C.** 2-methyl-2-methoxypentane

    B. 2-methylpent-3-ene   **D.** 2-methylpentene

8. Which of the following is the major product of this reaction?

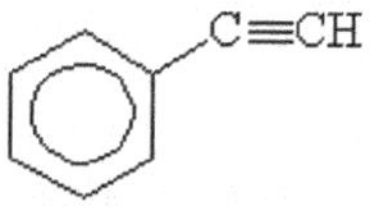

$$\text{C}_6\text{H}_5-\text{C}\equiv\text{CH} \quad + \quad 1)\ BH_3,\ THF \quad + \quad 2)\ {}^-OH,\ H_2O_2,\ H_2O \rightarrow$$

A. 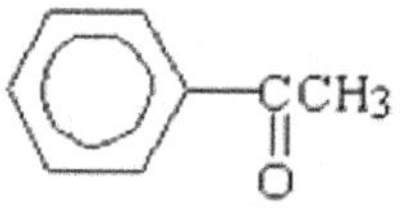 $C_6H_5-\overset{\displaystyle O}{\underset{\displaystyle \|}{C}}CH_3$ **C.** $C_6H_5-CH_2\overset{\displaystyle O}{\underset{\displaystyle \|}{C}}H$

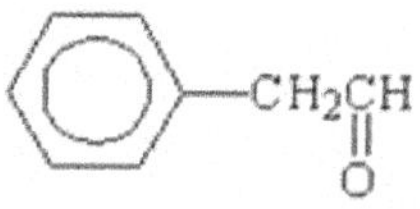

B. 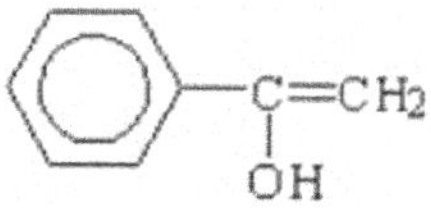 $C_6H_5-\overset{\displaystyle C}{\underset{\displaystyle OH}{}}=CH_2$ **D.** 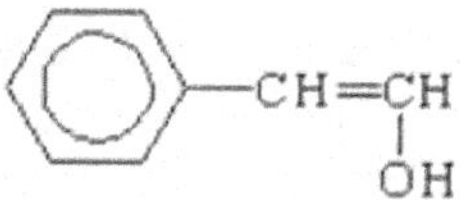 $C_6H_5-CH=\overset{\displaystyle CH}{\underset{\displaystyle OH}{}}$

9. Which of the following statements is supported by the table below?

	Solubility (g/L H_2O)	Melting point (°C)
para-nitrophenol	10	112
meta-nitrophenol	2.7	98
ortho-nitrophenol	0.8	47

    A. *Meta-* and *para*-nitrophenol form intramolecular hydrogen bonds

    B. *Ortho*-nitrophenol does not form intermolecular hydrogen bonds

    C. *Ortho*-nitrophenol has the greatest intramolecular hydrogen bonding

    D. *Para*-nitrophenol has the weakest intermolecular hydrogen bonding

10. When (*R*)-2-hexanol is subjected to PBr$_3$, the compound it produces is:

A. (*R*)-2-bromohexane

B. (*S*)-2-bromohexane

C. (*R/S*)-2-bromohexane

D. (*S*)-2-bromopentane

11. Which of the following are enol forms of 2-butanone (CH$_3$COCH$_2$CH$_3$)?

A. 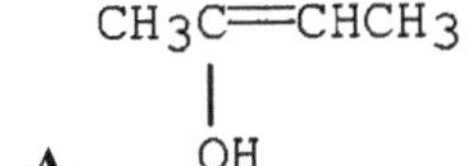 and 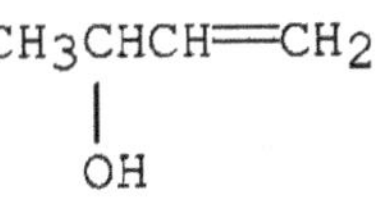 and 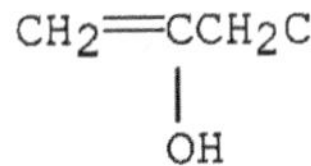

B. 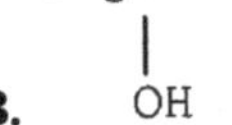 and

C. and

D. and

12. Which of the following acids is likely to have the weakest conjugate base?

A. CH$_3$Cl$_2$CCO$_2$H

B. CH$_3$CH$_2$CH$_2$CO$_2$H

C. (CH$_3$CH$_2$)$_3$CCO$_2$H

D. CH$_3$HNCH$_2$CH$_2$CH$_2$CO$_2$H

13. Which of the following products may be formed in the reaction below?

$+ \ H_2O \ / \ H_3O^+ \rightarrow \ ?$

A. HOCH$_2$CH(CH$_3$)$_2$

B. CH$_3$CH(CH$_3$)$_2$

C. HOOCCH$_2$CH(CH$_3$)$_2$

D. CH$_3$CH=CHCHO

14. Dodecylamine, CH$_3$(CH$_2$)$_{10}$CH$_2$NH$_2$, is insoluble in water but can be converted to a water-soluble form. Which of the species below represents a water-soluble form of this compound?

A. CH$_3$(CH$_2$)$_{10}$CH$_2$NHCH$_3$

B. CH$_3$(CH$_2$)$_{10}$CH$_2$NH$_3$$^+Cl^-$

C. CH$_3$(CH$_2$)$_{10}$CH$_2$NH$_2$–OH

D. CH$_3$(CH$_2$)$_{10}$CH$_2$NH$_2$–Cl

15. What is the name of the following compound?

$$H_3CCH_2CH_2CH_2CH_2CONH_2$$

A. 1-hexanamide

B. hexanamide

C. hexanamine

D. hexamine

16. Identify the correctly drawn arrows:

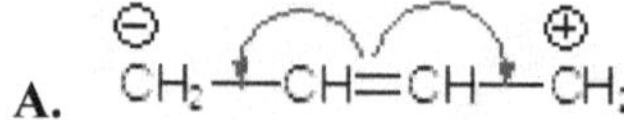

A.

B.

C.

D.

17. Which of the following describes the geometry about the carbon-carbon double bond in the alkene below?

A. *E*

B. *Z*

C. *cis*

D. *S*

18. The IR absorption at 1710 cm^{-1} most likely indicates which of the following functional groups?

A. carbon-carbon double bond

B. ketone

C. alcohol

D. ether

19. Which of the following laboratory techniques is most convenient for the separation of acetone from octane?

A. column chromatography

B. thin layer chromatography

C. distillation

D. crystallization

20. Which of the following is the best leaving group?

A. Cl$^-$

B. NH$_2{}^-$

C. $^-$OH

D. Br$^-$

21. What reagents can best be used to accomplish the following transformation?

A. 1) $Hg(OAc)_2$, H_2O/THF; 2) $NaBH_4$

B. 1) $Hg(O_2CCF_3)_2$, CH_3OH; 2) $NaBH_4$

C. 1) $BH_3 \cdot THF$; 2) HO^-, H_2O_2

D. H^+, H_2O

22. Which of the following statements correctly describes the general reactivity of alkynes?

A. Unlike alkenes, alkynes fail to undergo electrophilic addition reactions

B. Alkynes are generally more reactive than alkenes

C. An alkyne is an electron-rich molecule and therefore reacts as a nucleophile

D. The σ bonds of alkynes are higher in energy than the π bonds, and thus are more reactive

23. Which of the molecules shown below is NOT an *aromatic* compound?

A. 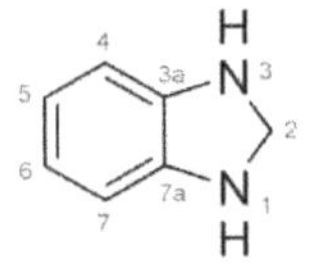 Benzimidazole

B.  Thiophene

C. 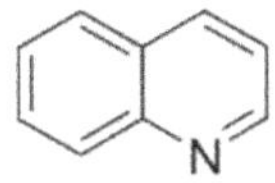 Quinoline

D. 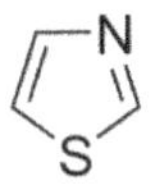 Thiazole

24. Which of the following reactions yields an ester?

A. $C_6H_5OH + CH_3CH_2Br$

B. $CH_3COOH + C_2H_5OH + H_2SO_4$

C. $CH_3COOH + SOCl_2$

D. $2\ CH_3OH + H_2SO_4$

25. Which statement is true regarding the major differences between aldehydes and ketones compared to other carbonyl compounds?

A. The carbonyl group carbon atom in aldehydes and ketones is bonded to atoms that do not attract electrons strongly

B. The polar carbon-oxygen bond in aldehydes and ketones is less reactive than the hydrocarbon portion of the molecule

C. The carbonyl carbon in aldehydes and ketones has bond angles of 120°, unlike the comparable bond angles in other carbonyl compounds

D. The molar masses in aldehydes and ketones are much less than in the other types of compounds

26. The boiling point of acetic acid is 119 °C, and methyl acetate is 57 °C. What is the reason that acetic acid boils at a much higher temperature than methyl acetate?

acetic acid   *methyl acetate*

A. molecular mass

B. presence of an ester linkage

C. hydrophobic interactions

D. hydrogen bonding

27. Which is the most reactive of the four derivatives of a carboxylic acid?

A. anhydride

B. acid bromide

C. amide

D. ester

28. Which of the following molecules represents a tertiary amine?

A.

B.

C.

D.

29. Which of the following is NOT a proper condensed structural formula for an alkane?

A. $CH_3CHCH_3CH_2CH_3$

B. $CH_3CH_2CH_2CH_2CH_3$

C. $CH_3CH_3CH_3$

D. $CH_3CH_2CH_2CH_3$

30. Triethylamine, $(CH_3CH_2)_3N$, is a molecule in which the nitrogen atom is [   ] hybridized, and the molecular shape is [   ].

A. sp^3 … trigonal pyramidal

B. sp^3 … tetrahedral

C. sp^2 … tetrahedral

D. sp^2 … trigonal planar

31. Which of the following is a structural isomer of 2-methylbutane?

A. *n*-propane

B. 2-methylpropane

C. *n*-butane

D. *n*-pentane

32. A compound with a broad, deep IR absorption at 3300 cm^{-1} indicates the presence of:

A. acyl halide

B. alcohol

C. alkene

D. ketone

33. Which of the following is an example of the termination step for a free-radical chain reaction?

A. Cl–Cl + hv → 2 Cl·

B. 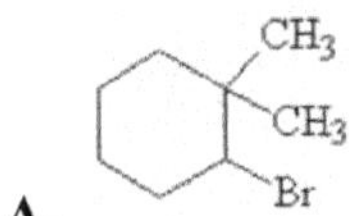

C. Cl· + Cl· → Cl$_2$

D.

34. Heating a(n) [   ] results in a Cope elimination.

A. amine oxide

B. imine

C. enamine

D. oxime

35. What is the major product when C$_2$H$_4$ undergoes an addition reaction with 1 mole equivalent of Br$_2$ in CCl$_4$?

A. C$_2$H$_2$ + H$_2$

B. C$_2$H$_4$Br$_2$

C. C$_2$HBr + HBr

D. C$_2$H$_3$Br

36. What is the degree of unsaturation for benzene?

A. 1

B. 2

C. 3

D. 4

37. What is the major product of the reaction of 2,2-dimethylcyclohexanol with HBr?

A.

B.

C.

D.

38. Which series of reactions best facilitates the following conversion?

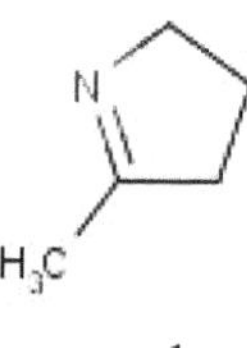

A. 1) $KMnO_4$ (*aq*); 2) $Hg(OAc)_2$ (*aq*); 3) $NaBH_4 / {}^-OH$

B. 1) $NaBH_4$; 2) H_3PO_4 / Δ

C. 1) H_3C-$MgBr$; 2) H_2O / H_3O^+

D. 1) $NaBH_4$; 2) HBr (*g*); 3) Mg / ether; 4) H_2O / H_3O^+

39. The pK_a of acetic acid (CH_3COOH) is 4.8. If the pH of an aqueous solution of CH_3COOH and CH_3COO^- is 4.8, then:

A. $[CH_3COOH] = [CH_3COO^-]$

B. $[CH_3COOH] < [CH_3COO^-]$

C. CH_3COOH is completely ionized

D. $[CH_3COOH] > [CH_3COO^-]$

40. What are the products of a hydrolysis reaction of an ester?

A. Alcohol and alkane

B. Carboxylic acid and alcohol

C. Ether and alkene

D. Ketone and aldehyde

41. Which of the following sequences ranks the following isomers in order of increasing boiling points?

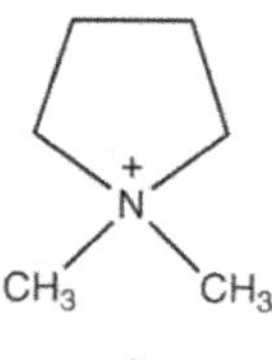

A. $1 < 3 < 2$ **B.** $3 < 2 < 1$ **C.** $2 < 1 < 3$ **D.** $2 < 3 < 1$

42. The name of the following alkyl group is $\sim CH_2CH_2CH_3$:

A. ethyl

B. propyl

C. isopropyl

D. *sec*-butyl

43. Which of the following is the most stable cation?

A. $CH_3\overset{\oplus}{C}=CH_2$

B. $CH_3CH=\overset{\oplus}{C}-$ (phenyl)

C. $CH_3-\overset{CH_3}{\underset{\oplus}{C}}-$ (phenyl)

D. $H_3C-\overset{CH_3}{\underset{\oplus}{C}}-CH_3$

44. Which of the following statements about *cis-* / *trans*-isomers is NOT correct?

    A. Conversion between *cis–* and *trans–* isomers occur by rotation around the double bond

    B. In the *trans–* isomer, the groups of interest are on opposite sides across the double bond

    C. In the *cis–* isomer, the reference groups are on the same side of the double bond

    D. There are no *cis–* / *trans–* isomers in alkynes

45. The compound $CH_3(CH_2)_5CH_3$ is:

    A. heptane   **C.** hexane

    B. hexene   **D.** pentane

46. Which of the following is NOT a resonance structure of the species shown?

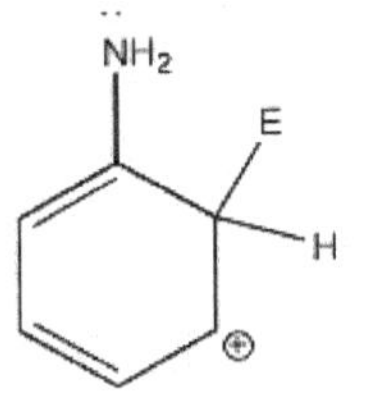

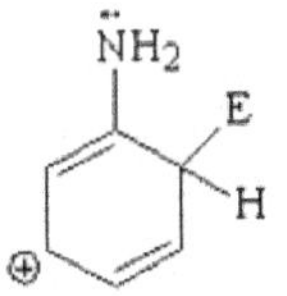

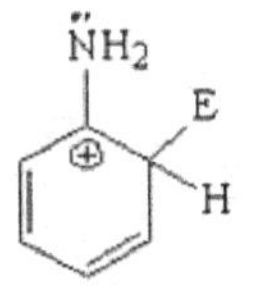

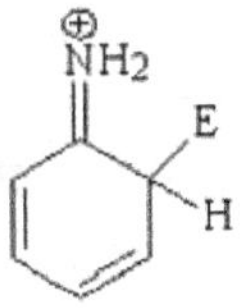

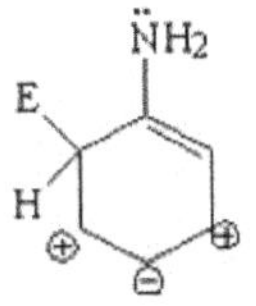

47. How many isomers are there of butane, C_4H_{10}?

    A. 3   **B.** 4   **C.** 1 (no isomers)   **D.** 2

48. Which of the following compounds gives the greatest number of proton NMR peaks?

    A. 3,3-dichloropentane   **C.** 1-chlorobutane

    B. 4,4-dichloroheptane   **D.** 1,4-dichlorobutane

49. After completing the synthesis of 3-methylpentanoic acid, which of the following treatments neutralizes the mineral acids and facilitates the organic acid distribution from the organic layer to the aqueous extraction layer?

A. Extraction with aqueous NaCl

B. Extraction with ether

C. Extraction with aqueous $NaHCO_3$

D. Extraction with water

50. Which of the following statements best describes the mechanism of the unimolecular elimination of *tert*-butyl chloride with ethanol?

A. The reaction is a concerted single-step process

B. The reaction involves homolytic cleavage of the C–Cl bond

C. The rate-determining step is the formation of $(CH_3)_3C\cdot$

D. The rate-determining step is the formation of $(CH_3)_3C^+$

51. Which of the following is NOT an example of a conjugated system?

A. 1,2-butadiene

B. cyclobutadiene

C. benzene

D. 2,4-cyclohexadiene

52. What reagents are used to convert 1-hexyne into 2-hexanone?

A. 1) Si_2BH; 2) H_2O_2, NaOH

B. Hg^{2+}, H_2SO_4, H_2O

C. 1) O_3; 2) $(CH_3)_2S$

D. 1) CH_3MgBr; 2) CO_2

53. What is the most likely regiochemistry of this electrophilic aromatic substitution reaction?

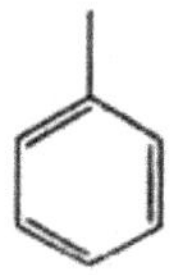

$$+ \; Br_2 / FeBr_3 \longrightarrow \; ?$$

A.

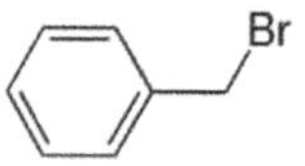

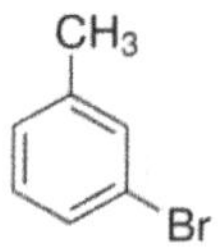

B.

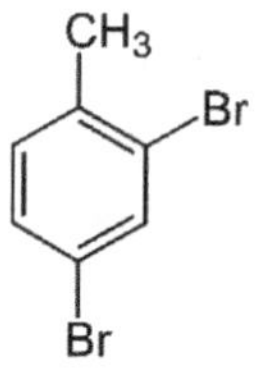

C.

D.

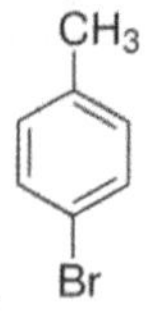

54. Compounds with the ~OH group attached to a saturated alkane-like carbon are:

A. ethers

B. hydroxyls

C. alcohols

D. alkyl halides

55. Which of the following reactions does NOT yield a ketone product?

A. 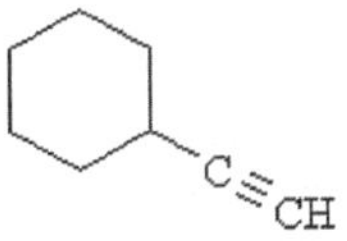 $C \equiv CH$ (cyclohexyl) $+ 1)\ Sia_2BH\,/\,THF;\ \ 2)\ H_2O_2\,/\,{}^-OH \rightarrow$

B. 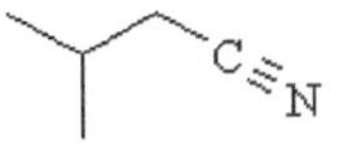 $C \equiv N$ (isobutyl) $+ 1)\ CH_3CH_2MgBr;\ \ 2)\ H_3O^+ \rightarrow$

C. 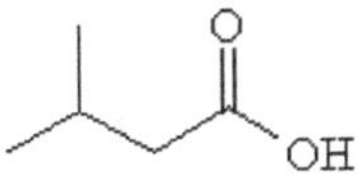 (3-oxobutanoic acid type, β-keto acid with OH) $+ 1)\ 2\ CH_3CH_2Li;\ \ 2)\ H_3O^+ \rightarrow$

D. 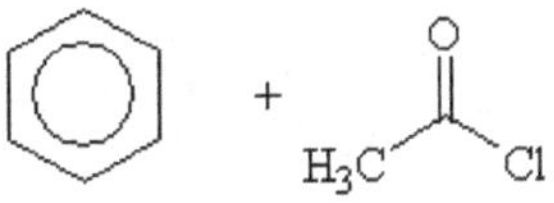 benzene $+\ H_3C\underset{Cl}{\overset{O}{\|}}C\ +\ AlCl3 \rightarrow$

56. Identify the carboxylic acid and alcohol from which the following ester was made.

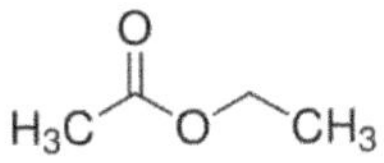

$H_3C\overset{O}{\overset{\|}{C}}O\text{---}CH_3$

A. $CH_3CH_2CO_2H$ and CH_3CH_2OH

B. CH_3CO_2H and CH_3CO_2H

C. CH_3CO_2H and CH_3CH_2OH

D. $CH_3CH_2CO_2H$ and $CH_3CH_2CH_2OH$

57. Identify the functional group:

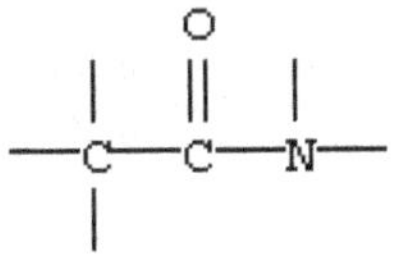

$$\text{---}\overset{}{\underset{}{C}}\text{---}\overset{O}{\overset{\|}{C}}\text{---}N\text{---}$$

A. anhydride

B. amine

C. amide

D. ester

58. Which of the following compounds has the lowest boiling point?

A. dimethylamine

B. *sec*-butylamine

C. diethylamine

D. *n*-butylamine

59. Name the following structure:

$$H_2C=CH-CH_2-C(CH_3)=CH_2$$

A. 2-methylene-4-pentene

B. 2-methyl-1,4-pentadiene

C. 2-methyl-2,4-pentadiene

D. 4-methyl-1,4-pentadiene

60. In an aqueous environment, which bond requires the most energy to break?

A. hydrogen

B. dipole-dipole

C. *sigma*

D. ionic

61. What is the relationship between the following compounds?

A. constitutional isomers

B. structural isomers

C. geometric isomers

D. conformational isomers

62. Where would one expect to find the ^{1}H NMR signal for the carboxyl group's hydrogen in propanoic acid?

A. δ 4.1-5.6 ppm

B. δ 10-13 ppm

C. δ 8-9 ppm

D. δ 6.1-7.8 ppm

63. Where in a petroleum fractionating tower would a chemist locate the molecules with the strongest intermolecular forces?

A. near the top

B. near the middle

C. near the bottom

D. no specific location because fractional distillation does not utilize molecular weight

64. Which of the following best describes the process of an S_N1 reaction in which the leaving group is on a chiral carbon atom?

A. inversion of stereochemistry

B. double inversion of stereochemistry

C. racemic mixture

D. retention of stereochemistry

65. Which of the following would show the LEAST regioselectivity for HBr addition?

A. $(CH_3)_2C=C(CH_3)CH_2CH_3$

B. $H_2C=C(CH_3)CH_2CH_3$

C. $CH_2HC=C(CH_3)CH_2CH_3$

D. $(CH_3)_2C=CHCH_2CH_3$

66. What is the major product of this reaction?

$$H_3C-C{\equiv}C-CH_3 + Na\ (s),\ NH_3\ (l) \rightarrow\ ?$$

A. (trans-2-butene structure)

B. (branched alkane structure)

C. (cis-2-butene structure)

D. (2-amino branched structure)

67. A compound is a six-carbon cyclic hydrocarbon. It is inert to bromine in water and bromine in dichloromethane, yet it decolorizes bromine in carbon tetrachloride when a small quantity of $FeBr_3$ is added. Which of the following is the identity of the compound?

A. 1,4-cyclohexadiene

B. 1,3-cyclohexadiene

C. benzene

D. cyclohexane

68. When (*S*)-2-heptanol is subjected to $SOCl_2$ / pyridine, the compound is transformed into:

A. (*R/S*)-2-chloroheptane

B. (*R*)-2-chloroheptane

C. (*S*)-2-chloroheptane

D. 2-heptone

69. What reagents are needed to complete the following synthesis?

$$R{-}\underset{\overset{\|}{O}}{C}{-}CH_3 \longrightarrow R{-}\underset{\overset{\|}{O}}{C}{-}O^-$$

A. 1) NaOH / heat; 2) HCl (*aq*)

B. 1) NaOH / Br_2

C. 1) warm conc. $KMnO_4$ / NaOH; 2) HCl (*aq*)

D. 1) Ag(NH_3)$_2$OH; 2) HCl (*aq*)

70. Which of the following has the highest boiling point?

A. ethyl alcohol, CH_3CH_2OH

B. acetic acid, CH_3COOH

C. ethane, CH_3CH_3

D. dimethyl ketone, CH_3COCH_3

Use the answer key to check your answers. Review the explanations in detail, focusing on questions you didn't answer correctly. Note the topic of those questions. Complete this BEFORE taking the next Diagnostic Test.

Notes for active learning

Diagnostic Test 2

This Diagnostic Test is designed to assess your proficiency on each topic and NOT to mimic the test. Use your test results and identify areas of strength and weakness to adjust your study plan and enhance your knowledge.

#	Answer:				Review	#	Answer:				Review
1:	A	B	C	D	___	36:	A	B	C	D	___
2:	A	B	C	D	___	37:	A	B	C	D	___
3:	A	B	C	D	___	38:	A	B	C	D	___
4:	A	B	C	D	___	39:	A	B	C	D	___
5:	A	B	C	D	___	40:	A	B	C	D	___
6:	A	B	C	D	___	41:	A	B	C	D	___
7:	A	B	C	D	___	42:	A	B	C	D	___
8:	A	B	C	D	___	43:	A	B	C	D	___
9:	A	B	C	D	___	44:	A	B	C	D	___
10:	A	B	C	D	___	45:	A	B	C	D	___
11:	A	B	C	D	___	46:	A	B	C	D	___
12:	A	B	C	D	___	47:	A	B	C	D	___
13:	A	B	C	D	___	48:	A	B	C	D	___
14:	A	B	C	D	___	49:	A	B	C	D	___
15:	A	B	C	D	___	50:	A	B	C	D	___
16:	A	B	C	D	___	51:	A	B	C	D	___
17:	A	B	C	D	___	52:	A	B	C	D	___
18:	A	B	C	D	___	53:	A	B	C	D	___
19:	A	B	C	D	___	54:	A	B	C	D	___
20:	A	B	C	D	___	55:	A	B	C	D	___
21:	A	B	C	D	___	56:	A	B	C	D	___
22:	A	B	C	D	___	57:	A	B	C	D	___
23:	A	B	C	D	___	58:	A	B	C	D	___
24:	A	B	C	D	___	59:	A	B	C	D	___
25:	A	B	C	D	___	60:	A	B	C	D	___
26:	A	B	C	D	___	61:	A	B	C	D	___
27:	A	B	C	D	___	62:	A	B	C	D	___
28:	A	B	C	D	___	63:	A	B	C	D	___
29:	A	B	C	D	___	64:	A	B	C	D	___
30:	A	B	C	D	___	65:	A	B	C	D	___
31:	A	B	C	D	___	66:	A	B	C	D	___
32:	A	B	C	D	___	67:	A	B	C	D	___
33:	A	B	C	D	___	68:	A	B	C	D	___
34:	A	B	C	D	___	69:	A	B	C	D	___
35:	A	B	C	D	___	70:	A	B	C	D	___

This page is intentionally left blank

1. Ethyl propanoate is a(n):

A. ester

B. alcohol

C. aldehyde

D. carboxyl alcohol

2. Acetone is a common solvent used in organic chemistry laboratories. Which of the following statements is/are correct regarding acetone?

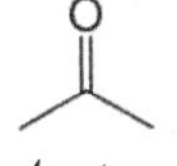

Acetone

   I. One atom is sp^3 hybridized and tetrahedral

   II. One atom is sp^2 hybridized and trigonal planar

   III. The carbonyl carbon contains an unshared pair of electrons

A. I only

B. II only

C. I and II only

D. II and III only

3. Which of the following statement(s) for the compound *meso*-tartaric acid is/are true?

   I. achiral

   II. the polarimeter reads a zero deflection

   III. racemic mixture

A. I only

B. II only

C. III only

D. I and II only

4. Which of the following compounds generates only one signal on its ^{1}H NMR spectrum?

A. *tert*-butyl alcohol

B. 1,2-dibromoethane

C. toluene

D. methanol

5. When placed in an electric field for separation by gel electrophoresis, which of the following molecules migrates toward the cathode at physiological pH 7.35?

A. $HOOCCHNH_2CH_2–S–S–CH_2CHNH_2COOH$

B. $HOCCH_2CH_2CHNH_2COOH$

C. $H_2NCH_2CHNH_2COOH$

D. $HN(CH_3)COOH$

6. Halogenation of alkanes proceeds by the mechanism shown below:

I. $Br_2 + h\nu \rightarrow 2\ Br\cdot$

II. $Br\cdot + RH \rightarrow HBr + R\cdot$

III. $R\cdot + Br_2 \rightarrow RBr + Br\cdot$

Which of these steps involve chain propagation?

A. I only

B. III only

C. I and II only

D. II and III only

7. When $CH_3\text{–}CH\text{=}CH_2$ is reacted in water with a catalytic amount of acid, a new compound is formed. What might be the product of this reaction?

A. $CH_3\text{–}\overset{\overset{O}{\|}}{C}\text{–}CH_3$ **B.** $H_3C\overset{OH}{\diagup}CH_3$ **C.**   **D.** $CH_3\text{–}CH\text{–}CH_2$

8. Which of the alkyne addition reactions below involve(s) an enol intermediate?

I. hydroboration/oxidation

II. treatment with $HgSO_4$ in dilute H_2SO_4

III. hydrogenation

A. I only

B. II only

C. III only

D. I and II only

9. If bromobenzene is treated with sulfur trioxide (SO_3) and concentrated sulfuric acid (H_2SO_4), what is/are the major product(s)?

A. *ortho-* and *para-*bromobenzenesulfonic acid   **C.** benzenesulfonic acid

B. benzene   **D.** *meta-*bromobenzenesulfonic acid

10. When (*R*)-2-heptanol is subjected to a two-step mechanism of tosyl chloride followed by Cl⁻, the product is:

A. (*R*)-2-chloroheptane   **C.** (*R/S*)-2-chloroheptane

B. heptene   **D.** (*S*)-2-chloroheptane

11. Of the following, which is the best solvent for an aldol addition?

A. acetone   **C.** propanol

B. methyl acetate   **D.** dimethyl ether

12. Which of the following are the preferred reagents used in the following synthesis?

A. 1) $NaBH_4$ / THF; 2) H_3O^+

B. 1) Mg / ether; 2) dry CO_2; 3) H_3O^+

C. 1) $LiAlH_4$ / THF; 2) H_3O^+

D. 1) Hot $KMnO_4$; 2) H_3O^+

13. What is the major ring-containing product of this reaction?

$+ \; KOH, \; H_2O / \Delta \rightarrow H^+ \; \text{workup} \rightarrow \; ?$

A.

B.

C.

D.

14. Which of the following amines is the most basic in the gas phase?

A. NH_3

B. $(CF_3)_3N$

C. $(CH_3)_3N$

D. H_2NCH_3

15. Provide the common name of the compound:

A. isoheptyl chloride

B. *tert*-heptyl chloride

C. neoheptyl chloride

D. *sec*-heptyl chloride

16. Which of the following molecules represents the most stable carbocation?

A. 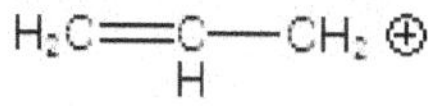 $H_2C=C(H)-CH_2^{\oplus}$

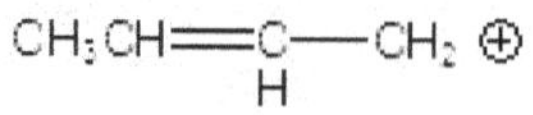

B. 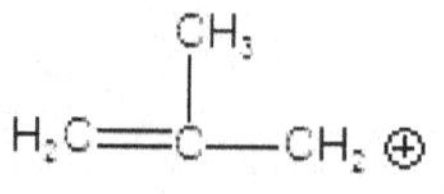 $H_2C=C(CH_3)-CH_2^{\oplus}$

C. $CH_3CH=C(H)-CH_2^{\oplus}$

D. 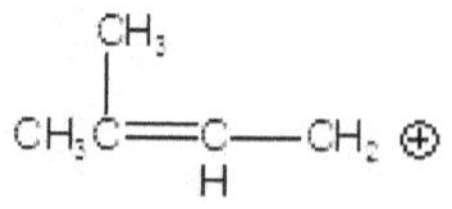 $CH_3C(CH_3)=C(H)-CH_2^{\oplus}$

17. Which of the following molecules are identical?

I., 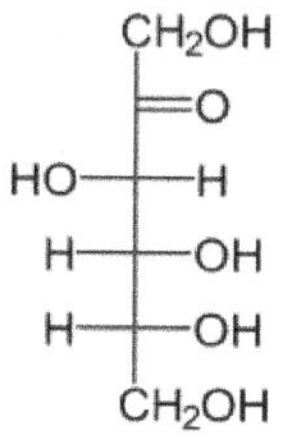 II., 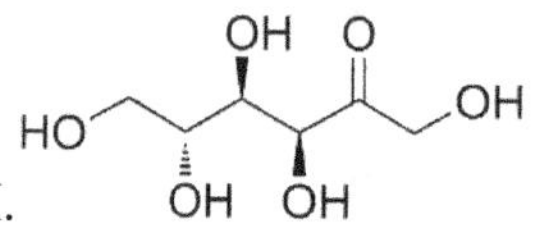 III.

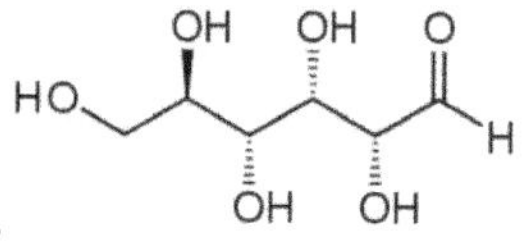

A. I and II

B. I and III

C. II and III

D. I, II and III

18. Methyl salicylate and aspirin exhibit a strong absorption for IR at 1735 cm^{-1}. This absorption indicates the presence of:

A. ester

B. aromatic ring

C. alcohol

D. phenol

19. Which of the following molecules are isolated closest to the top of a fractionating tower during distillation?

A. $C_{10}H_{22}$

B. $C_{20}H_{42}$

C. C_4H_{10}

D. C_8H_{18}

20. What statement(s) is/are true about an S_N2 reaction?

    I. A carbocation intermediate is formed

    II. The rate-determining step is bimolecular

    III. The mechanism has two steps

A. I only

B. II only

C. I and III only

D. II and III only

21. What is the product of the following reaction?

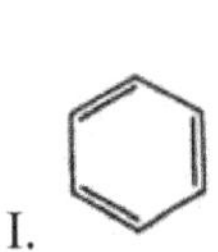

$+ \ H_2SO_4 / \Delta \rightarrow$?

A.   **B.**   **C.**   **D.**

22. Which of the following is the final and major product of this reaction?

$+ \ H_2O, \ H_2SO_4 / HgSO_4 \rightarrow$?

A.   **C.**

B.   **D.**

23. Rank the following three molecules in increasing order according to the rate they react with $Br_2 / FeBr_3$.

I.   II.   III.

A. II < III < I   **C.** I < II < III

B. II < I < III   **D.** I < III < II

24. Phenol exists predominantly:

  A. in the keto form because its keto tautomer is antiaromatic

  B. in the keto form because its keto tautomer is nonaromatic

  C. in the enol form because its keto tautomer is antiaromatic

  D. in the enol form because its keto tautomer is nonaromatic

25. All of the following statements about oxidation of carbonyls are true, EXCEPT:

A. Benedict's test involves the reduction of Cu^{2+}

B. Tollens' test involves the reduction of Ag^+

C. oxidation of primary alcohols produces secondary alcohols

D. oxidation of secondary alcohols produces ketones

26. Which of the following molecules is the strongest acid?

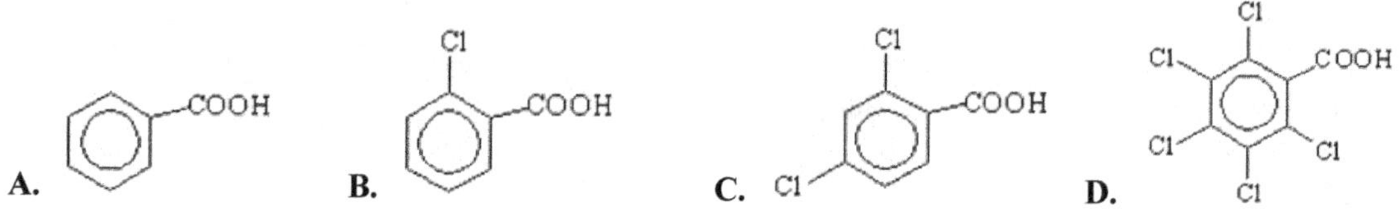

27. What is the name of the product formed by the reaction of butanoic acid with methylamine?

A. *N*-methylbutamide

B. *N*-methylbutanamide

C. pentanone

D. pentylamine

28. All of the following are properties of amines, EXCEPT:

A. amines react with acids to form amides at low temperatures

B. amines with low molecular weights are soluble in water

C. amines that form hydrogen bonds have higher boiling points relative to their molecular mass

D. amines frequently have offensive odors

29. What is the IUPAC name of the molecule shown?

A. 2-ethyl-5-hexene

B. 5-ethyl-1-hexene

C. 5-methyl-1-heptene

D. 3-methyl-6-heptene

30. Which of the following pairs are resonance structures?

31. Which of the following molecules contains a chiral carbon?

A. $CH_3-CH-CH_2CH_3$ with CH_3 below

C. $CH_3-C-CH_2CH_3$ with $=O$ below

B. $CH_3-CH-CH_2CH_3$ with OH below

D. $CH_3-CH-CH_3$ with OH below

32. Absorption of what type of electromagnetic radiation results in electronic transitions?

A. X-rays

B. radio waves

C. microwaves

D. ultraviolet light

33. Which of the following compounds undergo(es) a substitution reaction?

I. C_2H_6 II. C_2H_2 III. C_2H_4

A. I only

B. II only

C. I and II only

D. II and III only

34. Alkenes are more acidic than alkanes. What is the best explanation of this property?

A. The sp^2 hybridized orbitals in alkenes stabilize the negative charge generated when a proton is abstracted

B. The sp^2 hybridized orbitals in alkenes destabilize the negative charge generated when a proton is abstracted

C. The sp^3 hybridized orbitals in alkenes stabilize the negative charge generated when a proton is abstracted

D. The sp^3 hybridized orbitals in alkenes destabilize the negative charge generated when a proton is abstracted

35. If the compound C_5H_7NO contains 1 ring, how many *pi* bonds are there in this compound?

A. 0   **B.** 1   **C.** 2   **D.** 3

36. Which of the following is true about the benzene molecule?

A. It is a saturated hydrocarbon

B. The *pi* electrons of the ring move around the ring and have resonance

C. It is a hydrocarbon with the molecular formula of C_nH_{2n+2}

D. It contains heterocyclic oxygen

37. Treatment of salicylic acid with methanol and nonaqueous acid yields an:

$+ CH_3OH + H_2SO_4 \rightarrow ?$

Salicylic acid

A. acetal

B. ether

C. ester

D. hemiacetal

38. The reaction of ethylmagnesium bromide with which of the following compounds yields tertiary alcohol after quenching with aqueous acid?

A. ethylene oxide

B. $(CH_3)_2CO$

C. CH_3CHO

D. H_2CO

39. The water solubility of compounds containing the carboxylic acid group can be increased by reaction with:

A. water

B. sodium hydroxide

C. nitric acid

D. sulfuric acid

40. Amides are less basic than amines because the:

A. carbonyl group donates electrons by resonance

B. carbonyl group withdraws electrons by resonance

C. nitrogen does not have a lone pair of electrons

D. nitrogen has a full positive charge

41. Which of the compounds shown form hydrogen bonds between a mixture of the molecules?

A. $(CH_3)_3N$

B. $CH_3CH_2OCH_3$

C. $CH_3CH_2CH_2F$

D. $CH_3NHCH_2CH_3$

42. Which name is NOT correct?

A. 2,2-dimethylbutane

B. 2,3-dimethylpentane

C. 2,3,3-trimethylbutane

D. 2,3,3-trimethylpentane

43. Which of the following is closest to the C–O–C bond angle in CH_3–O–CH_3?

A. 109.5°

B. 90°

C. 180°

D. 120°

44. What is the relationship between the following compounds?

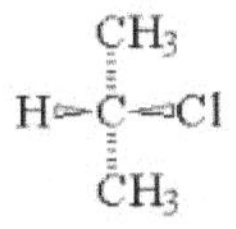

$$CH_3-\underset{CH_3}{\overset{CH_3}{H-C-Cl}} \quad \text{and} \quad$$

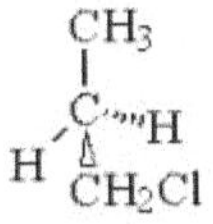

A. conformational isomers

B. constitutional isomers

C. diastereomers

D. enantiomers

45. Which of the following is a product of this reaction?

$$CH_3-\overset{O}{\overset{\|}{C}}-OCH_2CH_2CH_3 \ + \ NaOH \rightarrow \ ?$$

A. CH_3COOH

B. $CH_3CH_2COO^- \ Na^+$

C. CH_3CH_2-OH

D. $CH_3COO^- \ Na^+$

46. Which of these molecules is a tertiary amine?

A. RNH_2

B. R_2NH

C. R_3N

D. R_3NH^+

47. Name the structure:

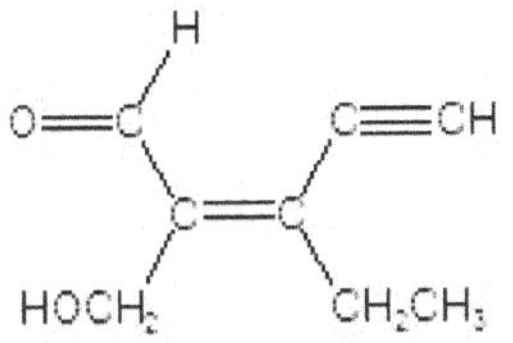

A. (*Z*)-3-ethyl-5-hydroxymethyl-3-penten-1-ynal

B. (*E*)-3-ethyl-5-hydroxymethyl-3-penten-1-ynal

C. (*Z*)-3-ethyl-2-hydroxymethyl-2-penten-4-ynal

D. (*E*)-3-ethyl-2-hydroxymethyl-2-penten-4-ynal

48. The nitrogen atom of trimethylamine is [   ] hybridized which is reflected in the C–N–C bond angle of [   ]:

A. *sp*, 180°

B. *sp*^{2}, 120°

C. *sp*^{2}, 108°

D. *sp*^{3}, 108°

49. Which of the following compounds are isomers?

I. $CH_3CH_2OCH_3$

II. $CH_3CH_2CH_2OH$

III. $CH_2COHCH_2CH_3$

IV. $CH_3CH_2OCH_2CH_3$

A. I and II only

B. II and III only

C. III and IV only

D. I, II and III only

50. While the carbonyl stretching frequency for simple aldehydes, ketones, and carboxylic acids is about 1710 cm^{-1}, the carbonyl stretching frequency for acid chlorides is about:

A. 1700 cm^{-1}

B. 1735 cm^{-1}

C. 1800 cm^{-1}

D. 1660 cm^{-1}

51. Which of the following is true regarding S_N1 and S_N2 reactions?

A. The rates of S_N1 reactions depend mainly on steric factors, while the rates of S_N2 reactions depend mainly on electronic factors

A. S_N1 reactions proceed more readily with a tertiary alkyl halide substrate, while S_N2 reactions proceed more readily with a primary alkyl halide substrate

C. S_N1 and S_N2 reactions produce rearrangement products

D. S_N1 reactions proceed via a carbocation intermediate, while S_N2 reactions proceed via a carbocation intermediate under certain conditions

52. Which statement is true in the oxymercuration-reduction of an alkene?

A. *Anti*-Markovnikov orientation and *anti*-addition occur

B. *Anti*-Markovnikov orientation and *syn*-addition occur

C. Markovnikov orientation and *anti*-addition occur

D. Markovnikov orientation and *syn*-addition occur

53. What is the major product of the following acid/catalyzed hydration reaction?

$+ H_2O / H_2SO_4, HgSO_4 \rightarrow$?

A.

B.

C.

D.

54. While electron-withdrawing groups (such as $-NO_2$ and $-CO_2R$) are *meta*-directing for electrophilic aromatic substitution reactions, they are *ortho-* / *para*-directing in nucleophilic aromatic substitution reactions. This observation would best be explained by using which concept?

A. tautomerism

B. aromaticity

C. hydrogen bonding

D. resonance

55. The ester prepared by heating 1-pentanol with acetic acid in the presence of an acidic catalyst is named:

A. 1-pentyl acetate

B. acetyl 1-pentanoate

C. acetic pentanoate

D. pentanoic acetate

56. A compound with an ~OH group and an ether-like –OR group bonded to the same carbon atom is:

A. hemiacetal

B. diol

C. aldol

D. acetal

57. The ion formed from a carboxylic acid is the:

A. ester cation

B. ester anion

C. carboxylate cation

D. carboxylate anion

58. Which sequence correctly ranks each carbonyl group in order of increasing reactivity toward nucleophilic addition?

A. $1 < 2 < 3$

B. $2 < 3 < 1$

C. $3 < 1 < 2$

D. $1 < 3 < 2$

59. Which formula best represents the form an amine takes in acidic solution?

A. RNH_2^-

B. RNH_2^+

C. RNH_2

D. RNH_3^+

60. What is the complete systematic IUPAC name for the following compound?

A. isopropyl-(4-isopropyl-4-methylbut-2-enyl) ether

B. (*E*)-4-isopropoxy-4,5-dimethylhex-2-ene

C. 4-(1-methylethoxy)-4-isopropyl-4-methylpent-2-ene

D. 4-isopropyl-2,4-dimethylhept-5-en-3-ol

61. Which of the following statements concerning the cyclic molecule shown is NOT true?

A. It contains a σ molecular orbital formed by the overlap of a carbon p atomic orbital with an oxygen sp^3 atomic orbital

B. It contains a σ molecular orbital formed by the overlap of carbon sp^3 hybrid atomic orbitals

C. It contains a σ molecular orbital formed by the overlap of carbon sp^2 hybrid atomic orbitals

D. It contains a π molecular orbital formed by the overlap of a carbon p atomic orbital with an oxygen p atomic orbital

62. The cause of *cis-trans* isomerism is:

A. short length of the double bond

B. strength of the double bond

C. lack of rotation of the double bond

D. stability of the double bond

63. Which of the following compounds is named correctly?

A. *meta*-fluorobenzoic acid

B. 2,5-dinitro-1-chlorobenzene

C. 2-iodo-1-bromobenzene

D. 1,3,dichloro-2-nitrobenzene

64. Which of the following is the most stable resonance contributor to acetic acid?

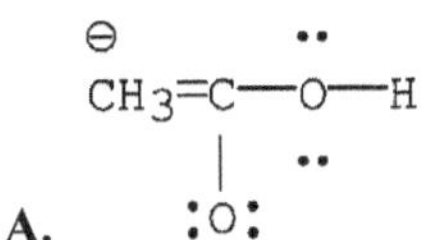

A.

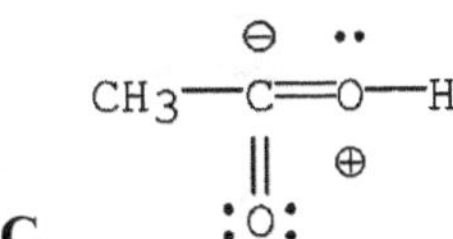

C.

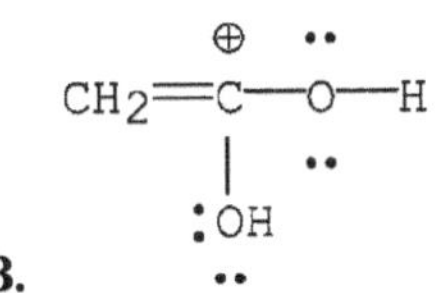

B.

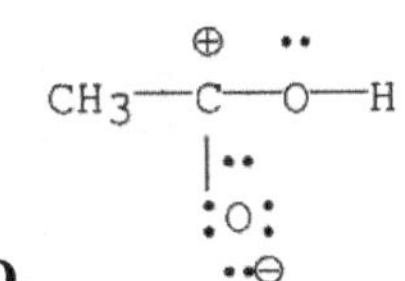

D.

65. $CH_3–CH_2–O–H$ and $CH_3–O–CH_3$ are a pair of compounds that are:

A. isomers

B. epimers

C. anomers

D. allotropes

66. A researcher recorded the NMR spectra of each of the following compounds. Disregarding chemical shifts, which molecule below possesses a spectrum significantly different from the others?

A. 1,1,2-tribromobutane

B. bromobutane

C. 3,3-dibromoheptane

D. dibutyl ether

67. Various hydrocarbons can be separated from a crude oil mixture by:

A. precipitation

B. fractional distillation

C. filtration

D. chromatography

68. The rate of an S_N2 reaction depends on:

A. neither the concentration of the nucleophile nor the substrate

B. the concentration of the nucleophile and the substrate

C. the concentration of the substrate only

D. the concentration of the nucleophile only

69. Which of the following has the lowest heat of hydrogenation per mole of H_2 absorbed?

A. 1,2-hexadiene

B. 1,3,5-heptatriene

C. 1,3-hexadiene

D. 1,5-hexadiene

70. Given that 1-butyne has a boiling point of 8.1 °C, what is the phase of propyne at room temperature and 1 atm pressure?

A. solid

B. supercritical fluid

C. gas

D. liquid

> Use the answer key to check your answers. Review the explanations in detail, focusing on questions you didn't answer correctly. Note the topic of those questions. Complete this BEFORE taking the next Diagnostic Test.

Notes for active learning

Notes for active learning

Diagnostic Test 3

This Diagnostic Test is designed to assess your proficiency on each topic and NOT to mimic the test. Use your test results and identify areas of strength and weakness to adjust your study plan and enhance your knowledge.

#	Answer:				Review	#	Answer:				Review
1:	A	B	C	D	___	36:	A	B	C	D	___
2:	A	B	C	D	___	37:	A	B	C	D	___
3:	A	B	C	D	___	38:	A	B	C	D	___
4:	A	B	C	D	___	39:	A	B	C	D	___
5:	A	B	C	D	___	40:	A	B	C	D	___
6:	A	B	C	D	___	41:	A	B	C	D	___
7:	A	B	C	D	___	42:	A	B	C	D	___
8:	A	B	C	D	___	43:	A	B	C	D	___
9:	A	B	C	D	___	44:	A	B	C	D	___
10:	A	B	C	D	___	45:	A	B	C	D	___
11:	A	B	C	D	___	46:	A	B	C	D	___
12:	A	B	C	D	___	47:	A	B	C	D	___
13:	A	B	C	D	___	48:	A	B	C	D	___
14:	A	B	C	D	___	49:	A	B	C	D	___
15:	A	B	C	D	___	50:	A	B	C	D	___
16:	A	B	C	D	___	51:	A	B	C	D	___
17:	A	B	C	D	___	52:	A	B	C	D	___
18:	A	B	C	D	___	53:	A	B	C	D	___
19:	A	B	C	D	___	54:	A	B	C	D	___
20:	A	B	C	D	___	55:	A	B	C	D	___
21:	A	B	C	D	___	56:	A	B	C	D	___
22:	A	B	C	D	___	57:	A	B	C	D	___
23:	A	B	C	D	___	58:	A	B	C	D	___
24:	A	B	C	D	___	59:	A	B	C	D	___
25:	A	B	C	D	___	60:	A	B	C	D	___
26:	A	B	C	D	___	61:	A	B	C	D	___
27:	A	B	C	D	___	62:	A	B	C	D	___
28:	A	B	C	D	___	63:	A	B	C	D	___
29:	A	B	C	D	___	64:	A	B	C	D	___
30:	A	B	C	D	___	65:	A	B	C	D	___
31:	A	B	C	D	___	66:	A	B	C	D	___
32:	A	B	C	D	___	67:	A	B	C	D	___
33:	A	B	C	D	___	68:	A	B	C	D	___
34:	A	B	C	D	___	69:	A	B	C	D	___
35:	A	B	C	D	___	70:	A	B	C	D	___

This page is intentionally left blank

1. Which of the following is *cis*-2,3-dichloro-2-butene?

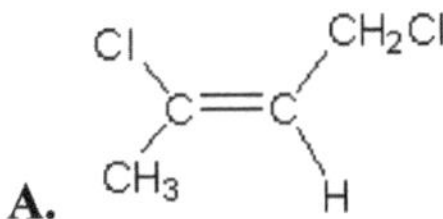

2. Which of the following compounds exhibits the greatest dipole moment?

A. (1*S*,2*S*)-1,2-dichloro-1,2-diphenylethane

B. 1,2-dichlorobutane

C. (1*R*,2*S*)-1,2-dichloro-1,2-diphenylethane

D. (*E*)-1,2-dichlorobutene

3. Which of the following is a result of the reaction below?

(*S*)-3-bromo-3-methylhexane + HCN

A. loss of optical activity

B. mutarotation

C. retention of optical activity

D. inversion of absolute configuration

4. What type of spectroscopy would be the LEAST useful in distinguishing dimethyl ether from bromoethane?

A. UV spectroscopy

B. mass spectrometry

C. IR spectroscopy

D. ^{1}H NMR spectroscopy

5. At atmospheric pressure, a particular organic liquid has a boiling point of 180 °C and decomposes at 170 °C, while its isomer, also a liquid, boils at 220 °C and decomposes at 195 °C. These isomers can be separated by:

A. sublimation

B. fractional distillation

C. vacuum distillation

D. simple distillation

6. Which of the following is the most stable conformer of *trans*-1-isopropyl-3-methylcyclohexane?

A. methyl and isopropyl are axial

B. methyl and isopropyl are equatorial

C. methyl is axial and isopropyl is equatorial

D. methyl is equatorial and isopropyl is axial

7. The reaction below can be classified as a(n):

cis-pent-2-ene → pentane

A. tautomerization

B. elimination

C. oxidation

D. reduction

8. What is the major product of this reaction?

$+ CH_3CH_2MgBr \rightarrow$

A. 

C. 

B. 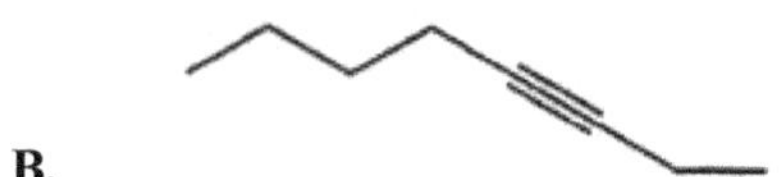

D. 

9. The reason why complete hydrogenation of benzene to cyclohexane requires H_2, rhodium (Rh) catalyst, and 1,000 psi pressure at 100 °C is because:

A. the double bonds in benzene have the same reactivity as *pi* bonds of non-aromatic alkenes

B. hydrogenation produces an aromatic compound

C. the double bonds in benzene are more reactive than a typical alkene

D. the double bonds in benzene are less reactive than a typical alkene

10. Which of the following molecules has the most acidic proton?

A. 2-pentanol

B. 3-pentyne

C. 2-pentene

D. pentane

11. Oxidation of an aldehyde produces a:

A. tertiary alcohol

B. secondary alcohol

C. primary alcohol

D. carboxylic acid

12. Which of the following reactions will NOT result in the formation of a carboxylic acid?

A. Oxidation of a secondary alcohol

B. Oxidation of a primary alcohol

C. Hydrolysis of nitriles

D. Grignard reagents reacting with CO_2

13. Which of the following compounds is the most susceptible to nucleophilic attack by ⁻OH?

A. propionyl bromide

B. benzyl bromide

C. propanal

D. butanoic acid

14. What is the conjugate acid of CH_3NH_2?

A. NH_4^+

B. NH_2^-

C. $CH_3NH_3^+$

D. CH_3NH^-

15. Which of the following is the IUPAC name for this compound?

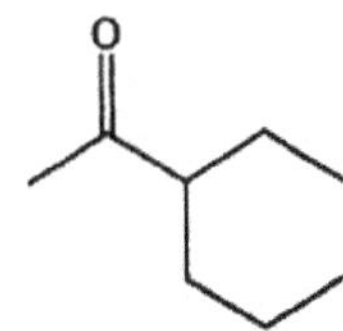

A. 3-ethylhexan-2-one

B. 4-ethylhexan-5-one

C. 3-propylpentan-2-one

D. 3-propylpentan-4-one

16. Draw a structural formula for benzene. How many σ bonds are in the molecule?

A. 14   **B.** 18   **C.** 6   **D.** 12

17. If two chlorine atoms replace two hydrogen atoms in butane, how many different dichlorobutane constitutional isomers can there be?

A. 4   **B.** 6   **C.** 2   **D.** 1

18. What 1H NMR spectral data is expected for the compound shown?

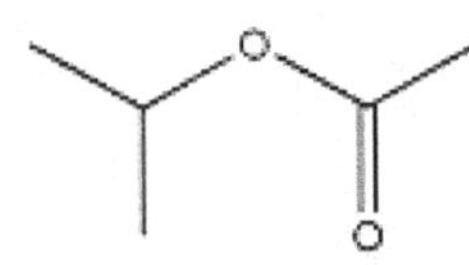

A. 4.9 (1H, sextet), 4.3 (3H, singlet), 3.0 (6H, doublet)

B. 3.6 (3H, singlet), 2.8 (3H, septet), 1.2 (6H, doublet)

C. 4.3 (1H, septet), 3.3 (3H, singlet), 1.2 (6H, doublet)

D. 3.8 (1H, septet), 2.2 (3H, singlet), 1.0 (6H, doublet)

19. A compound which dissolves in hydrochloric acid, and not in neutral water, is:

A. 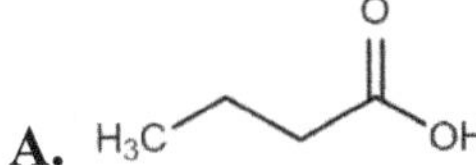

B. $H_3C–CH_2–CH_2–NH_2$

C. 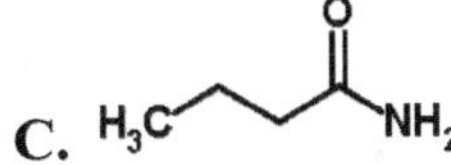

D. $H_3C–CH_2–O–H$

20. Which of the following compounds readily undergoes E_1, S_N1 and E_2 reactions, but not S_N2 reactions?

A. $(CH_3CH_2CH_2)_3CBr$

B. $CH_3CH_2CH_2CH_3$

C. $(CH_3CH_2)_3COH$

D. $CH_3CH_2CH_2CH_2Br$

21. The major product of the following reaction will likely be the result of a(n) [   ] mechanism?

2-bromobutane + *tert*-butyl alkoxide

A. E_2 **B.** E_1 **C.** S_N2 **D.** S_N1

22. Which of the following is the product for the reaction?

$$HC\equiv C-CH_2CH_2-CH_3 + H_2O + HgSO_4 / H_2SO_4 \rightarrow$$

A. $CH_3CH_2CH_2C(OH)=CH_2$

B. $CH_3CH_2CH_2CH=CHOH$

C. $H_3C-CH_2CH_2-C(=O)-CH_3$

D. $CH_3CH_2CH_2CH_2CHO$

23. Which of the following reactions is NOT an electrophilic aromatic substitution reaction?

A. $CH_3C_6H_5 + C_6H_5CH_2CH_2Cl / AlCl_3$

B. $CH_3C_6H_5 + Br_2 / FeBr_3$

C. $CH_3C_6H_5 + CH_3CH_2CH_2COCl / AlCl_3$

D. $CH_3C_6H_5 + H_2, Rh / C$

24. What is the major product of this reaction?

$$CH_3CH_2CH_2CH_2OH + \text{(pyridinium } ClCrO_3^-) \rightarrow \; ?$$

A. (butanal)

B. (butanoic acid)

C. (1-butene)

D. (1-chlorobutane)

25. Which of the following will alkylate a lithium enolate most rapidly?

A. methyl bromide

B. isopropyl bromide

C. neopentyl bromide

D. bromobenzene

26. Which acid would be expected to have the lowest boiling point?

A. oxalic

B. formic

C. benzoic

D. acetic

27. What is the major organic product of this reaction?

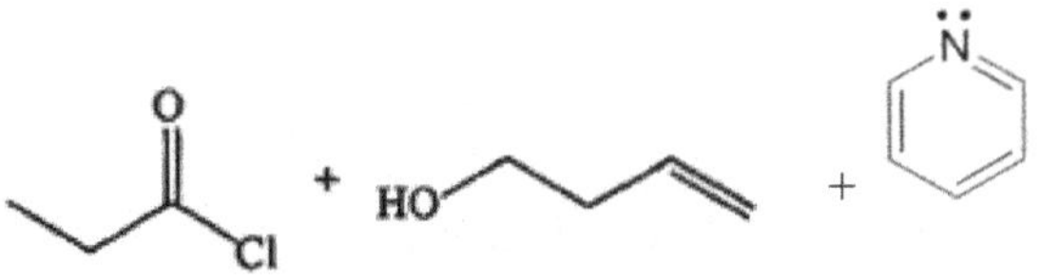

A.

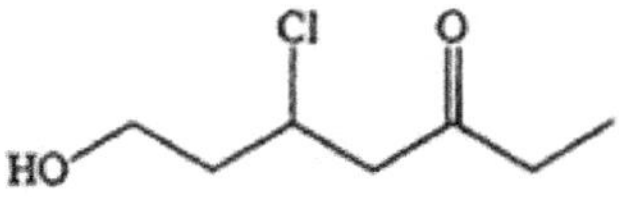

B.

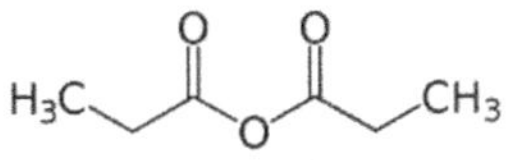

C.

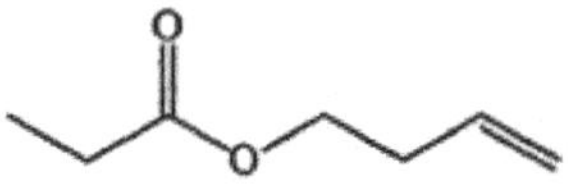

D.

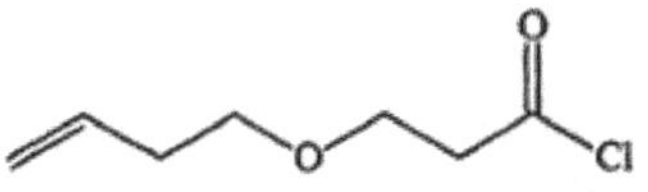

28. Which amine has the lowest boiling point?

A.

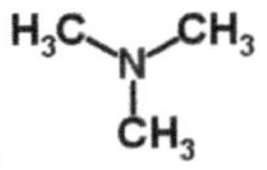

B.

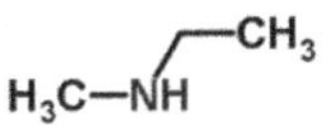

C.

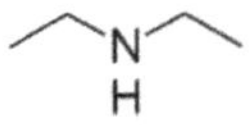

D.

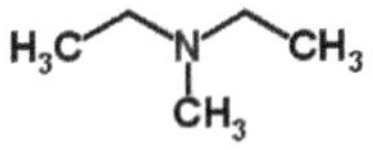

29. Name the structure shown below:

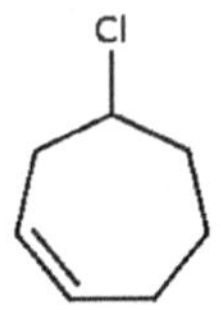

A. 1-chloro-3-cycloheptene

B. 4-chlorocycloheptene

C. 4-chlorocyclohexene

D. 1-chloro-3-cyclohexene

30. Which is the formal charge of nitrogen in NH_4?

A. –2   **B.** –1   **C.** 0   **D.** +1

31. How many asymmetric centers are present in the compound shown below?

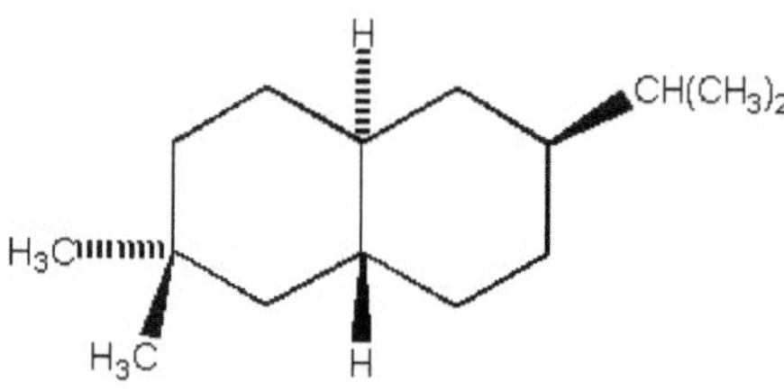

A. 1   **B.** 2   **C.** 3   **D.** 4

32. Which of the following compounds absorbs the longest wavelength of UV-visible light?

A. (Z)-1,3-hexadiene

B. (E)-1,3,5-hexatriene

C. (Z)-but-2-ene

D. (E)-but-2-ene

33. The complete combustion of one mole of nonane in oxygen would produce [ ] moles of CO_2 and [ ] moles of H_2O?

A. 9 … 10   **B.** 9 … 9   **C.** 9 … 4.5   **D.** 4.5 … 4.5

34. What is the major product from the following reaction?

+ HBr → ?

A. (image: epoxide with CH₂Br) —CH₂Br

B. Me (image: Br, OH)

C. HO (image: structure with Br)

D. H_3C (image: ketone) CH_3

35. What is the product from the reaction of one mole of acetylene and two moles of bromine vapor?

A. 1,1,2,2-tetrabromoethene

B. 1,1,2,2-tetrabromoethane

C. 1,2-dibromoethene

D. 1,2-dibromoethane

36. Which two steps may be used to synthesize 1-chloro-4-nitrobenzene, starting from benzene?

A. 1) Na / NH_3; 2) Cl_2 / $FeCl_3$

B. 1) HNO_3 / H_2SO_4; 2) Cl_2 / $FeCl_3$

C. 1) HCl / H_2O; 2) HNO_3 / H_2SO_4

D. 1) Cl_2 / $FeCl_3$; 2) HNO_3 / H_2SO_4

37. Which of the following alcohols dehydrates with the fastest rate?

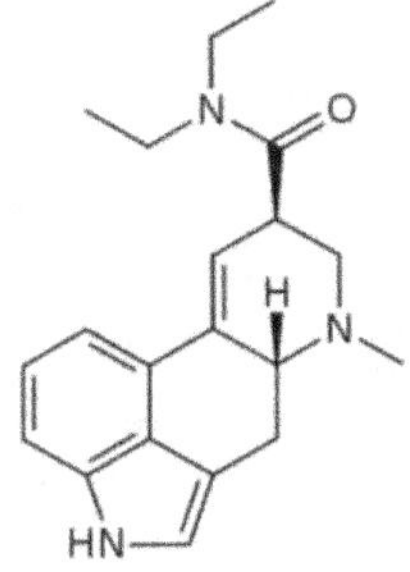

38. In Benedict's test:

    I. an aldehyde is oxidized

    II. a red/brown precipitate is formed

    III. copper (II) ion is reduced

A. I only

B. II only

C. III only

D. I, II and III

39. The reaction of butanoic acid with ethanol produces:

A. butyl ethanamide

B. ethyl butanamide

C. butyl ethanoate

D. ethyl butanoate

40. What reagent(s) are needed to complete the following reaction?

A. H_2, Pd

B. 1) DIBAL–H; 2) H_3O^+

C. 1) $LiAlH_4$; 2) H_3O^+

D. 1) $NaBH_4$; 2) H_3O^+

41. In water, does the molecule lysergic acid diethylamide function as:

    I. acid   II. base   III. neither acid nor base

A. I only

B. II only

C. III only

D. I and II only

42. Provide the IUPAC name of the compound:

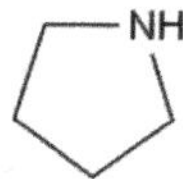

A. 1,1-dimethyl-5-chloropentane

B. 6-chloro-2-methylhexane

C. 1-chloro-5-methylhexane

D. 2-methyl-chloro-heptane

43. The nitrogen's lone pair in pyrrolidine is occupying which type of orbital?

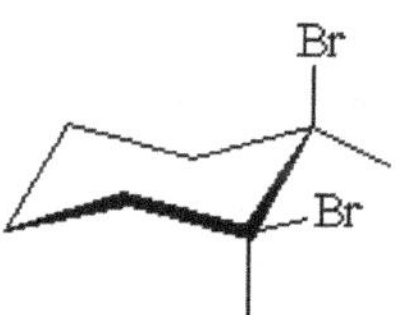

A. s **B.** sp^3 **C.** sp^2 **D.** sp

44. What is the relationship between the structures shown below?

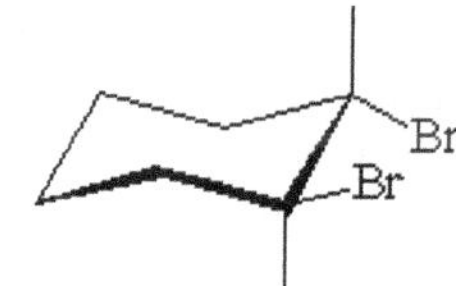

A. identical compounds

B. conformational isomers

C. diastereomers

D. enantiomers

45. In the electrophilic aromatic substitution of phenol, substituents add predominantly in which position(s)?

    I. *ortho* to the hydroxyl group

    II. *meta* to the hydroxyl group

    III. *para* to the hydroxyl group

A. I only

B. II only

C. III only

D. I and III only

46. The compound below has which functional groups?

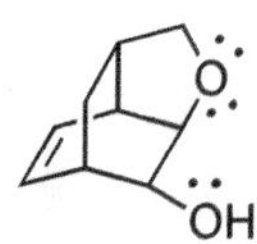

A. ether, alkene and alcohol

B. ester, alkene and alcohol

C. aromatic, alcohol and ether

D. aromatic, alcohol and ester

47. Which observation denotes a positive Benedict's test?

   A. A mirror-like deposit forms from a colorless solution

   B. A purple solution yields a brown precipitate

   C. A red precipitate forms from a blue solution

   D. A red-brown solution becomes clear and colorless

48. Which fatty acid is expected to have the highest boiling point?

   A. oxalic, $(CO_2H)_2$ **C.** benzoic, $C_6H_5CO_2H$

   B. stearic, $CH_3(CH_2)_{16}CO_2H$ **D.** acetic, CH_3CO_2H

49. Esters and amides are most easily made by nucleophilic acyl substitution reactions on:

   A. acid chlorides   **C.** carboxylates

   B. acid anhydrides   **D.** carboxylic acids

50. Which of the compounds listed is the strongest organic base that functions as a proton acceptor?

   A.   **C.** $CH_3{-}CH_2{-}NH_2$

   B.   **D.**

51. Name the compound shown below.

   A. *cis*-1,3-dichlorocyclohexane   **C.** *cis*-1,2-dichlorocyclohexane

   B. *trans*-1,3-dichlorocyclohexane   **D.** *trans*-1,2-dichlorocyclohexane

52. The compound methylamine, CH_3NH_2, contains a C–N bond. In this bond, which of the following best describes the charge on the nitrogen atom?

   A. uncharged   **C.** +1

   B. slightly negative   **D.** slightly positive

53. What is the relationship between the two structures shown below?

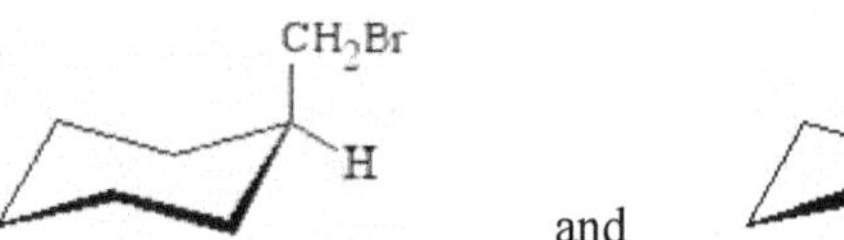

and

A. conformational isomers

B. configurational isomers

C. enantiomers

D. constitutional isomers

54. Which compound(s) show(s) intense IR absorption at 1680 cm^{-1}?

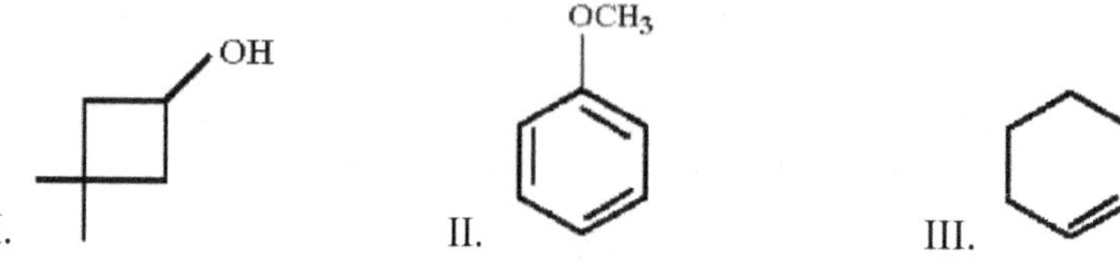

A. I only

B. II only

C. III only

D. II and III only

55. Which one of the species below is soluble in dilute HCl (*aq*)?

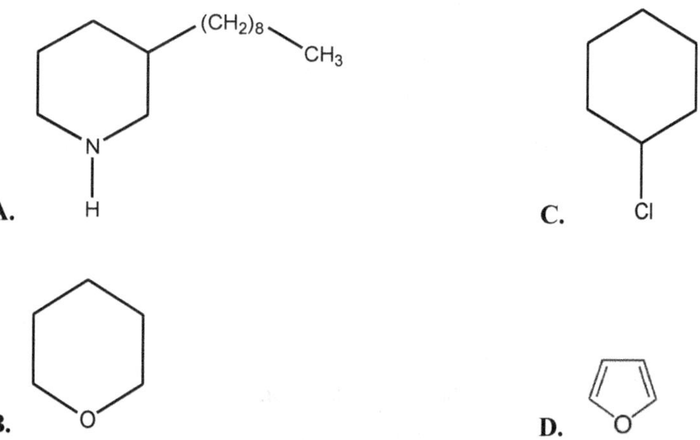

56. The heat of combustion of propane is –530 kcal/mol. A reasonable approximation for the heat of combustion for heptane is:

A. –432 kcal/mol

B. –682 kcal/mol

C. –865 kcal/mol

D. –1158 kcal/mol

57. Which of the following is the best solvent for the addition of HCl to 3-hexene?

A. 3-hexene

B. CH$_3$COOH

C. CH$_3$OH

D. H$_2$O

58. How many moles of hydrogen are required to convert a mole of pentyne to pentane?

A. 0   **B.** 1   **C.** 2   **D.** 3

59. Which of the following is NOT aromatic?

A. 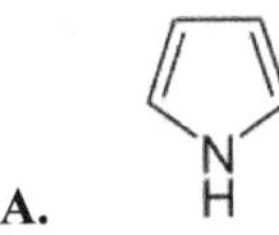   **C.** 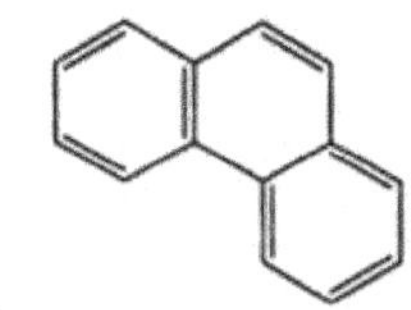

B. 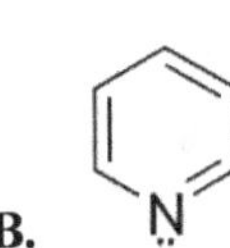   **D.**

60. Which compound is NOT an unsaturated compound?

A. $H_2C{=}CH{-}Cl$ **C.** $H_2C{=}CH_2$
B. $CH_3{-}CH_2{-}O{-}H$ **D.** $CH_3{-}CH{=}CH_2$

61. A ylide is a molecule that can be described as a:

A. carbanion bound to a positively charged heteroatom
B. carbocation bound to a positively charged heteroatom
C. carbocation bound to a carbon radical
D. carbocation bound to a diazonium ion

62. Which formula correctly illustrates the form taken by the acetic acid in a basic solution?

63. The reaction of an amine and a carboxylic acid produces what kind of compound?

A. amine   **C.** anhydride
B. amide   **D.** ketone

64. Which of the following behaves like a base?

I. $CH_3CONHCH_3$ II. $(CH_3)_2NH$ III. $C_2H_5CONHCH_3$

A. I only

B. II only

C. III only

D. I and II only

65. What is the IUPAC name of the compound shown?

$$CH_3-CH-CH_2-CH-CH_3$$

with CH_3 below the first CH and OH below the second CH.

A. 2,2-dimethyl-4-butanol

B. 4,4-dimethyl-2-butanol

C. 2-methyl-4-pentanol

D. 4-methyl-2-pentanol

66. Which of the following molecules is the most polar?

A. acetaldehyde

B. acetic acid

C. ethane

D. ethylene

67. Among the butane conformers, which occur at energy minima on a graph of potential energy versus dihedral angle?

A. *anti*

B. eclipsed

C. *gauche*

D. eclipsed and *gauche*

68. What is the relationship between H_a and H_b in the following structure?

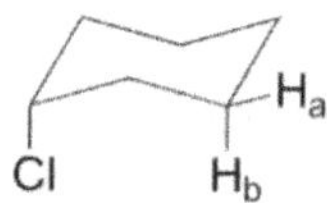

A. diastereotopic

B. enantiotopic

C. homotopic

D. isotopic

69. Which compound is least soluble in water?

A. 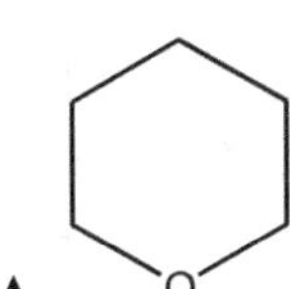

C.

B.

D.

70. What is the major product of the following reaction?

$$+ \ H^+ \ / \ H_2O \rightarrow \ ?$$

A. CH_2OH

C. CH_3, OH

B. OH, CH_3

D. CH_3, OH

Use the answer key to check your answers. Review the explanations in detail, focusing on questions you didn't answer correctly. Note the topic of those questions. Complete this BEFORE taking the next Diagnostic Test.

Notes for active learning

Diagnostic Test 4

This Diagnostic Test is designed to assess your proficiency on each topic and NOT to mimic the test. Use your test results and identify areas of strength and weakness to adjust your study plan and enhance your knowledge.

#	Answer:				Review	#	Answer:				Review
1:	A	B	C	D	___	36:	A	B	C	D	___
2:	A	B	C	D	___	37:	A	B	C	D	___
3:	A	B	C	D	___	38:	A	B	C	D	___
4:	A	B	C	D	___	39:	A	B	C	D	___
5:	A	B	C	D	___	40:	A	B	C	D	___
6:	A	B	C	D	___	41:	A	B	C	D	___
7:	A	B	C	D	___	42:	A	B	C	D	___
8:	A	B	C	D	___	43:	A	B	C	D	___
9:	A	B	C	D	___	44:	A	B	C	D	___
10:	A	B	C	D	___	45:	A	B	C	D	___
11:	A	B	C	D	___	46:	A	B	C	D	___
12:	A	B	C	D	___	47:	A	B	C	D	___
13:	A	B	C	D	___	48:	A	B	C	D	___
14:	A	B	C	D	___	49:	A	B	C	D	___
15:	A	B	C	D	___	50:	A	B	C	D	___
16:	A	B	C	D	___	51:	A	B	C	D	___
17:	A	B	C	D	___	52:	A	B	C	D	___
18:	A	B	C	D	___	53:	A	B	C	D	___
19:	A	B	C	D	___	54:	A	B	C	D	___
20:	A	B	C	D	___	55:	A	B	C	D	___
21:	A	B	C	D	___	56:	A	B	C	D	___
22:	A	B	C	D	___	57:	A	B	C	D	___
23:	A	B	C	D	___	58:	A	B	C	D	___
24:	A	B	C	D	___	59:	A	B	C	D	___
25:	A	B	C	D	___	60:	A	B	C	D	___
26:	A	B	C	D	___	61:	A	B	C	D	___
27:	A	B	C	D	___	62:	A	B	C	D	___
28:	A	B	C	D	___	63:	A	B	C	D	___
29:	A	B	C	D	___	64:	A	B	C	D	___
30:	A	B	C	D	___	65:	A	B	C	D	___
31:	A	B	C	D	___	66:	A	B	C	D	___
32:	A	B	C	D	___	67:	A	B	C	D	___
33:	A	B	C	D	___	68:	A	B	C	D	___
34:	A	B	C	D	___	69:	A	B	C	D	___
35:	A	B	C	D	___	70:	A	B	C	D	___

This page is intentionally left blank

1. What is the IUPAC name for the following structure?

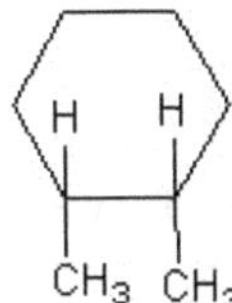

A. 5,6-dimethyl cyclohexane

B. *cis*-1,2-dimethyl cyclohexane

C. *trans*-1,2-dimethyl cyclohexane

D. 1,2-dimethyl cyclohexane

2. $C=C$, $C=O$, $C=N$ and $N=N$ bonds are observed in many organic compounds. However, $C=S$, $C=P$, $C=Si$, and other similar bonds are not often found. What is the most probable explanation for this observation?

A. the comparative sizes of $3p$ atomic orbitals make effective overlap between them less likely than between two $2p$ orbitals

B. S, P and Si do not undergo hybridization of orbitals

C. S, P and Si do not form π bonds due to the lack of occupied p orbitals in their ground state electron configurations

D. carbon does not combine with elements below the second row of the periodic table

3. Which of the following compounds share the same absolute configuration?

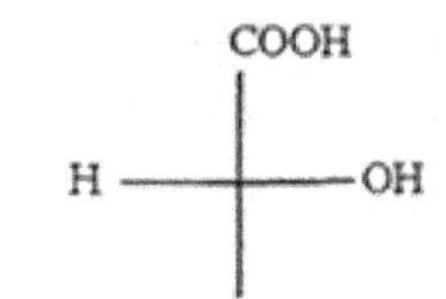

I.

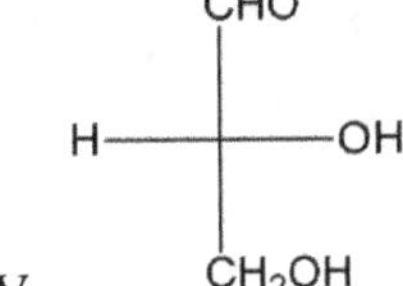

III.

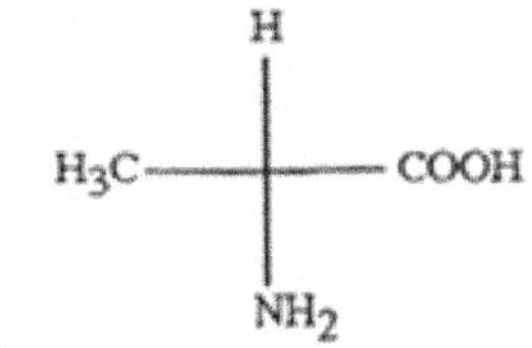

II.

IV.

A. I and III only

B. I and II only

C. I and IV only

D. II, III and IV only

4. Which of the following compounds is NOT IR active?

A. Cl_2

B. CO

C. $CH_3CH_2CH_2OH$

D. CH_3Br

5. Which of the following techniques would be most suitable for separating and analyzing two miscible liquids with boiling points of 60 °C and 140 °C?

    A. Simple distillation and gas chromatography   **C.** Recrystallization and electrophoresis

    B. Vacuum distillation and electrophoresis   **D.** Recrystallization and gas chromatography

6. Ethers can be formed from ethyl bromide in a reaction whereby the incoming $^-$OR group represents a(n):

    A. substrate   **C.** nucleophile

    B. electrophile   **D.** leaving group

7. Products A, B, and C of the following reaction are, respectively:

2-methyl-2-butene + HBr	$\rightarrow$	product A
2-methyl-2-butene + HBr / H_2O_2	$\rightarrow$	product B
2-methyl-2-butene + H_2O / H^+ / heat	$\rightarrow$	product C

	Product A	Product B	Product C
A.	2-bromo-2-methylbutane	2-bromo-3-methylbutane	2-methyl-2-butanol
B.	3-methyl-2-bromobutene	3-methyl-2-bromobutane	2-methyl-2-butanol
C.	2-bromo-2-methylbutane	2-bromo-2-methylbutane	3-methyl-2-butanol
D.	1-bromo-2-methyl-2-butene	2-bromo-2-methylbutane	2-methyl-2-butane

8. Which is the most stable product for the reaction below:

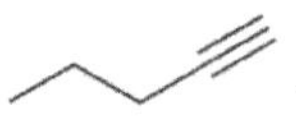

    + 1) BH_3 / THF 2) $^-$OH, H_2O_2, H_2O

    A. 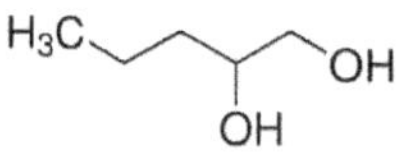 H_3C—CH(OH)—CH(OH) structure

    C. H_3C—C(=O)—CH$_2$—CH$_3$ (2-butanone type)

    B. $CH_3CH_2CH_2CH{=}CHOH$

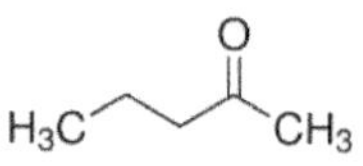

    D. 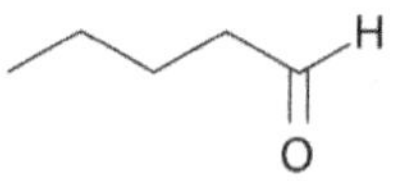 pentanal structure

9. Which of the following compounds is least susceptible to electrophilic aromatic substitution?

    A. p-H_3CCH_2O–C_6H_4–O–CH_2CH_3 **C.** p-Cl–C_6H_4–$NH_3{}^+$

    B. p-O_2N–C_6H_4–NH–CH_3 **D.** p-CH_3CH_2–C_6H_4–CH_2CH_3

10. Which of the following would have the highest boiling point?

A. 1-hexyne

B. 1-hexene

C. hexane

D. 1-hexanol

11. All of the statements concerning the carbonyl group in aldehydes and ketones are true, EXCEPT:

A. in condensed form, the aldehyde group can be written as –CHO

B. since the bond is polar, carbonyl groups readily form hydrogen bonds with each other

C. the bond angles about the central carbon atom are 120°

D. the bond is polar with a slight negative charge on the oxygen atom

12. Which of the following molecules would be expected to be the most soluble in water?

A. $CH_3(CH_2)_6CO_2H$

B. $CH_3(CH_2)_{12}CO_2H$

C. CH_3CO_2H

D. $CH_3CH_2CH_2CO_2H$

13. Which of the following type of bond is depicted below?

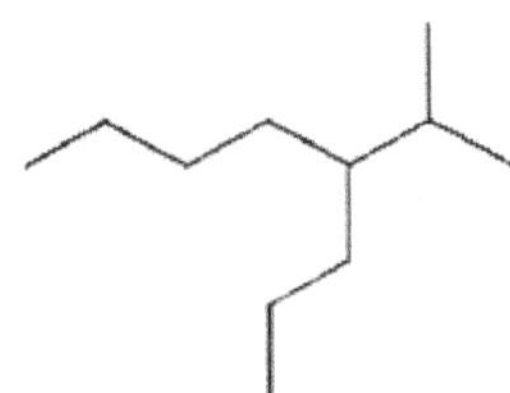

A. amide bond

B. glycosidic bond

C. ester bond

D. ether bond

14. What is the conjugate base of CH_3NH_2?

A. NH_2^-

B. NH_4^+

C. CH_3NH^-

D. $CH_3NH_3^+$

15. Give the IUPAC name for the following structure:

A. 4-isopropyloctane

B. 5-isopropyloctane

C. 3-ethyl-2-methylheptane

D. 2-methyl-3-ethylheptane

16. How many σ bonds are in cyclohexane (C_6H_{12}), a saturated cyclic hydrocarbon?

A. 12   **B.** 14   **C.** 16   **D.** 18

17. What is the correct IUPAC name for the following structure?

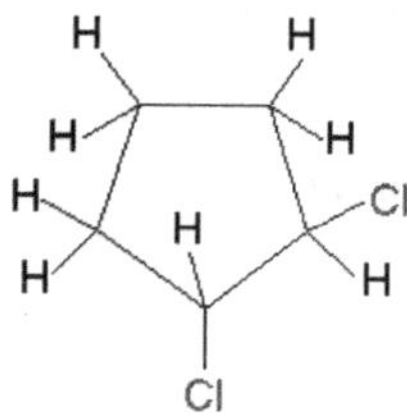

A. *trans*-1,2-dichlorocyclopentane

B. 1,2-dichlorocyclopentane

C. *cis*-1,2-dichlorocyclopentane

D. *trans*-dichlorocyclopentane

18. Which sequence correctly ranks the regions of the electromagnetic spectrum in order of increasing energy?

    1) infrared   2) ultraviolet   3) radio wave

A. $3 < 1 < 2$ **C.** $2 < 1 < 3$

B. $3 < 2 < 1$ **D.** $1 < 3 < 2$

19. The solvent diethyl ether can be mixed with water but only by shaking the two liquids together. After the shaking is stopped, the liquids separate into two layers. The protonated form of the alkaloid caffeine is readily soluble in water but not in diethyl ether. Suggest what may happen to the caffeine of a caffeinated beverage if the beverage is first made alkaline with sodium hydroxide and then shaken with diethyl ether.

A. The diethyl ether and water would mix into one layer

B. The caffeine would transform into the free acid and transfer into the diethyl ether

C. The caffeine would transform into the free base and become more soluble in the diethyl ether

D. The water layer would turn a pink color, indicating an alkaline pH

20. Which of the following compounds can most easily undergo both E_1 and S_N2 reactions?

A. $(CH_3CH_2CH_2)_2CHBr$ **C.** $(CH_3CH_2CH_2)_3CCH_2Cl$

B. $(CH_3CH_2CH_2)_3CBr$ **D.** $(CH_3CH_2CH_2)_2CHCN$

21. Which of the following is vinyl chloride?

A. $CH_2=CHCl$ **B.** $CH_2=CHCH_2Cl$ **C.**   **D.**

22. The compound 1-butyne contains:

   A. a ring structure

   B. a triple bond

   C. a double bond

   D. all single bonds

23. Which of the following statements is NOT correct about benzene?

   A. The carbon-carbon bond lengths are the same

   B. The carbon-hydrogen bond lengths are the same

   C. The carbon atoms are *sp* hybridized

   D. It has delocalized electrons

24. The functional group C=O is in all the species below, EXCEPT:

   A. amides

   B. ethers

   C. aldehydes

   D. ketones

25. Which of the following compound is most soluble in water?

   A. acetone

   B. cyclohexanone

   C. 2-butanone

   D. 3-butanone

26. Which of the following molecule is the most polar?

   A. butane

   B. butanoic acid

   C. cyclohexane

   D. ethanol

27. The compound below has which functional groups?

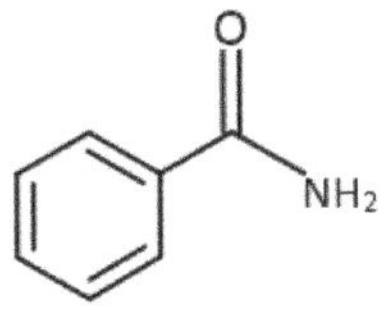

   A. aromatic and alcohol

   B. aromatic and amine

   C. aromatic and amide

   D. aromatic and carboxylic acid

28. A functional group containing nitrogen is in:

   A. carboxylic acids

   B. amines

   C. alcohols

   D. alkenes

29. Which name is NOT correct for IUPAC nomenclature?

A. 2,2-dimethylbutane

B. 2,3-dimethylpentane

C. 2,3,3-trimethylbutane

D. 2,3,4-trimethylpentane

30. Triethylamine [$(CH_3CH_2)_3N$] is a molecule in which the nitrogen atom is [ ] hybridized, and the C–N–C bond angle is approximately [ ].

A. sp^3, > 109.5°

B. sp^3, 109.5°

C. sp^2, > 109.5°

D. sp^2, < 109.5°

31. Which of the following compounds is NOT chiral?

A. 1,2-dichlorobutane

B. 1,4-dibromobutane

C. 2,3-dibromobutane

D. 1,3-dibromobutane

32. In the UV-visible spectrum of (*E*)-1,3,5-hexatriene, the lowest energy absorption corresponds to a:

A. π to σ^* transition

B. σ to σ^* transition

C. σ to π transition

D. π to π^* transition

33. Which of the following molecules has the lowest boiling point?

A. *cis*-2-pentene

B. 2-pentyne

C. pentane

D. neopentane

34. What reagent(s) is/are needed to accomplish the following transformation?

A. BH_3 / THF

B. 1) BH_3 / THF 2) ^-OH, H_2O_2, H_2O

C. H_2O / H_2O_2

D. H_2O / H^+

35. The *pi* bond of an alkyne is [ ] and [ ] than the *pi* bond of an alkene.

A. longer; stronger

B. longer; weaker

C. shorter; stronger

D. shorter; weaker

36. Which reaction is NOT characteristic of aromatic compounds?

A. addition

B. halogenation

C. nitration

D. sulfonation

37. Which of the following is an allylic alcohol?

A. $CH_3CH=CHCH_2OH$

B. $HOCH=CHCH_2CH_3$

C. $CH_2=CHCH_2CH_3$

D. $CH_2=CHCH_2OCH_3$

38. Treatment of a nitrile with a Grignard reagent, followed by hydrolysis, results in:

A. ester

B. ketone

C. aldehyde

D. ether

39. The most common reactions of carboxylic acid or its derivative involve:

A. replacement of the group bonded to the carbonyl atom

B. oxidation of the R group

C. addition across the double bond between carbon and oxygen

D. replacement of the oxygen atom in the carbonyl group

40. What are the products of an acid-catalyzed hydrolysis reaction of amides and water?

A. an alcohol and an alkane

B. an amine and a ketone

C. a carboxylic acid and an amine salt

D. an ester and an ether

41. Which of the following is NOT correct about amines?

A. Amines are bases (proton acceptors)

B. Amines are converted to ammonium salts by reaction with HCl

C. Amines are organic derivatives of ammonia

D. Amines are very water-soluble

42. Which compound has the common name *sec*-butylamine?

A. 1-butanamine

B. 2-butanamine

C. *N*-methyl-1-propanamine

D. *N*-methyl-2-propanamine

43. How many distinct and degenerate p orbitals exist in the second electron shell, where n = 2?

A. 3   **B.** 2   **C.** 1   **D.** 0

44. The specific rotation of a pure enantiomeric substance is −6.30°. What is the percentage of this enantiomer in a mixture with an observed specific rotation of −3.15°?

A. 75%   **B.** 80%   **C.** 25%   **D.** 50%

45. What is the product of the reaction of one mole of acetylene and one mole of bromine vapor?

A. 1,1,2,2-tetrabromoethane

B. 1,1,2,2-tetrabromoethene

C. 1,2-dibromoethane

D. 1,2-dibromoethene

46. Which of these molecules is aromatic?

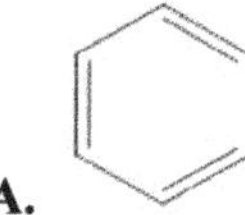

A.

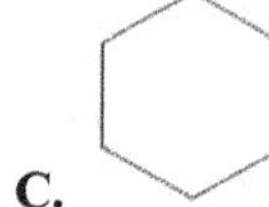

C.

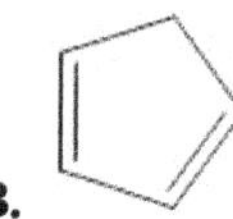

B.

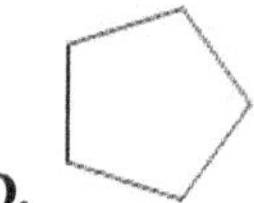

D.

47. Compounds of the type R_3C-OH are referred to as [   ] alcohols.

A. primary

B. secondary

C. tertiary

D. quaternary

48. Which of the following represents the correct ranking in terms of increasing the boiling point?

A. *n*-butane < 1-butanol < diethyl ether < 2-butanone

B. *n*-butane < 2-butanone < diethyl ether < 1-butanol

C. 2-butanone < *n*-butane < diethyl ether < 1-butanol

D. *n*-butane < diethyl ether < 2-butanone < 1-butanol

49. When alcohol reacts with a carboxylic acid, the major product is a(n):

A. salt

B. ester

C. amine

D. amide

50. Penicillin can be taken orally if the compound can survive degradation by the low pH of the stomach. The amide bond between the *R* group and 6-APA of dicloxacillin is stable at low pH because:

A. high hydroxide concentration promotes amide stability

B. amide bonds do not easily undergo hydrolysis

C. hydrogen bonding has stabilization effects

D. aromatic compounds are unaffected by changes in pH

51. The reaction of an amine with water is best represented by:

A. $R-NH_2 + H_2O \leftrightarrow R-NH_3^+ + OH^-$

B. $R-NH_2 + 2\,H_2O \leftrightarrow R-NH_4^{2+} + 2\,OH^-$

C. $R-NH_2 + 2\,H_2O \leftrightarrow R-N^{2-} + 2$

D. $R-NH_2 + H_2O \leftrightarrow R-NH^- + H_3O^+$

52. What is the IUPAC name of the compound shown below?

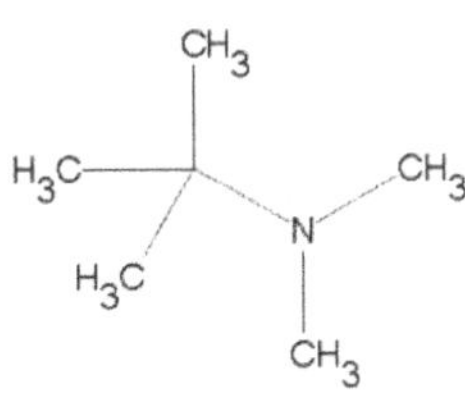

- **A.** (1*R*,4*S*)-1,4-dichloro-1-ethyl-4-methylcyclopentane
- **B.** (1*R*,3*S*)-1,3-dichloro-1-ethyl-3-methylcyclopentane
- **C.** (1*S*,3*S*)-1,3-dichloro-1-ethyl-3-methylcyclopentane
- **D.** (1*R*,3*S*)-1,3-dichloro-1-methyl-3-ethylcyclopentane

53. A molecule of acetylene (C_2H_2) has a [   ] geometry and a molecular dipole moment that is [   ].

- **A.** bent, zero
- **B.** linear, nonzero
- **C.** linear, zero
- **D.** bent, nonzero

54. How many different isomers are there for dibromobenzene, $C_6H_4Br_2$?

- **A.** 0
- **B.** 1
- **C.** 2
- **D.** 3

55. Provide the IUPAC name of the following compound:

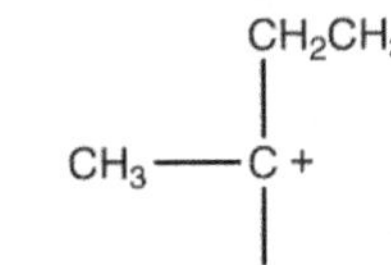

- **A.** *N,N*,2-trimethyl-2-propanamine
- **B.** *N,N*,2-trimethylpropanamine
- **C.** *N,N*,1,1-tetramethylethanamine
- **D.** *N,N*-dimethyl-2-butanamine

56. Identify the most stable carbocation:

A. $H_2C{=}CH \oplus$

C. 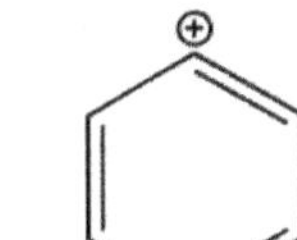

B.

D.

57. When two compounds are made up of the same number and kind of atom but differ in their molecular structure, they are:

A. hydrocarbons

B. isomers

C. homologs

D. isotopes

58. Which NMR signal represents the most deshielded proton?

A. δ 2.0

B. δ 3.8

C. δ 6.5

D. δ 7.3

59. The temperatures in a fractionating tower at an oil refinery are important, but so are the pressures. Where might the pressure in a fractional distillation tower be greatest, at the bottom or the top, and why?

A. At the bottom, because the lower temperature means a greater number of vaporized molecules

B. At the top, because the lower temperature means a greater number of vaporized molecules

C. At the bottom, because the higher temperature means a greater number of vaporized molecules

D. At the top, because the higher temperature means a greater number of vaporized molecules

60. An alkyl halide forms a carbocation that is more stable than the carbocation formed from isopropyl bromide. Which of the following alkyl halides forms the most stable carbocation?

A. *n*-propyl chloride

B. *tert*-butyl chloride

C. methyl chloride

D. ethyl chloride

61. The rate law for the addition of HBr to many simple alkenes may be approximated as rate = k [alkene]·[HBr]. This rate law indicates all the following EXCEPT that the reaction:

A. occurs in a single step involving one HBr molecule and one alkene molecule

B. involves one HBr molecule and one alkene molecule in the rate-determining step and involves many steps

C. is first-order in HBr

D. is second-order overall

62. Which of the following does NOT correctly describe the physical properties of an alkyne?

A. Less dense than water

B. Insoluble in most organic solvents

C. Relatively nonpolar

D. Nearly insoluble in water

63. Of the following, which reacts most readily with $Br_2 / FeBr_3$ in an electrophilic aromatic substitution?

A. 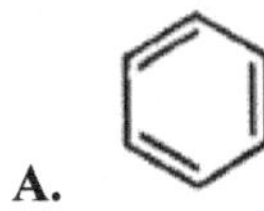

C. 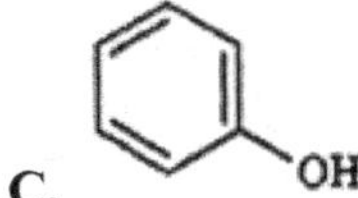

B. 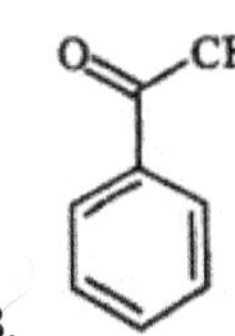

D. 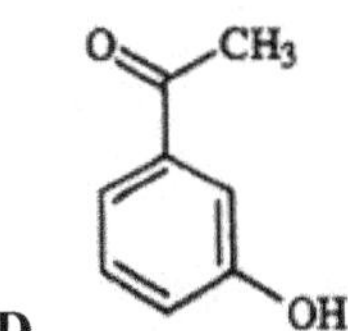

64. The functional group, ~OH, is in which one of these types of organic compounds?

A. amines

B. alcohols

C. alkanes

D. alkenes

65. Which of the following organic compounds is most likely NOT a ketone?

A. estradiol (female hormone)

B. progesterone (female hormone)

C. androsterone (male hormone)

D. cortisone (adrenal hormone)

66. Which acid is expected to have the highest boiling point?

A. formic

B. oxalic

C. acetic

D. stearic

67. What are the major organic products of the reaction shown?

$$+ H_2O / H_2SO_4 \rightarrow \,?$$

A. $CH_3COOH + HOCH_2CH_2CH_3$

B. $CH_3CH_2CH_2OH + CH_3CH_2OH$

C. $CH_3CH_2COOH + HOCH_2CH_3$

D. $CH_3CH_2CH_2COO^- + {}^+H_2OCH_2CH_3$

68. *p*-toluidine is somewhat soluble in water due to the polarity of:

A. its aromatic ring structure

B. its amine group

C. *p*-toludinoic acid

D. benzene

69. Name the following structure:

$$CH_3\text{-}C(CH_2CH_3)=C(CH_3)\text{-}CH_2CH_2CH_2OH$$

A. *cis*-3,4-dimethyl-3-hepten-7-ol

B. *trans*-4,5-dimethyl-4-hepten-1-ol

C. *cis*-4,5-dimethyl-4-hepten-1-ol

D. *trans*-3,4-dimethyl-3-hepten-7-ol

70. Consider the interaction of two hydrogen $1s$ atomic orbitals of the same phase. Which of the statements below is NOT an accurate description of this interaction?

A. The molecular orbital formed is cylindrically symmetric

B. The molecular orbital formed has a node between the atoms

C. The molecular orbital formed is lower in energy than a hydrogen $1s$ atomic orbital

D. A *sigma* bonding molecular orbital is formed

Use the answer key to check your answers. Review the explanations in detail, focusing on questions you didn't answer correctly. Note the topic of those questions. Complete this BEFORE taking the next Diagnostic Test.

Notes for active learning

Notes for active learning

Diagnostic Test 5

This Diagnostic Test is designed to assess your proficiency on each topic and NOT to mimic the test. Use your test results and identify areas of strength and weakness to adjust your study plan and enhance your knowledge.

#	Answer:				Review	#	Answer:				Review
1:	A	B	C	D	___	36:	A	B	C	D	___
2:	A	B	C	D	___	37:	A	B	C	D	___
3:	A	B	C	D	___	38:	A	B	C	D	___
4:	A	B	C	D	___	39:	A	B	C	D	___
5:	A	B	C	D	___	40:	A	B	C	D	___
6:	A	B	C	D	___	41:	A	B	C	D	___
7:	A	B	C	D	___	42:	A	B	C	D	___
8:	A	B	C	D	___	43:	A	B	C	D	___
9:	A	B	C	D	___	44:	A	B	C	D	___
10:	A	B	C	D	___	45:	A	B	C	D	___
11:	A	B	C	D	___	46:	A	B	C	D	___
12:	A	B	C	D	___	47:	A	B	C	D	___
13:	A	B	C	D	___	48:	A	B	C	D	___
14:	A	B	C	D	___	49:	A	B	C	D	___
15:	A	B	C	D	___	50:	A	B	C	D	___
16:	A	B	C	D	___	51:	A	B	C	D	___
17:	A	B	C	D	___	52:	A	B	C	D	___
18:	A	B	C	D	___	53:	A	B	C	D	___
19:	A	B	C	D	___	54:	A	B	C	D	___
20:	A	B	C	D	___	55:	A	B	C	D	___
21:	A	B	C	D	___	56:	A	B	C	D	___
22:	A	B	C	D	___	57:	A	B	C	D	___
23:	A	B	C	D	___	58:	A	B	C	D	___
24:	A	B	C	D	___	59:	A	B	C	D	___
25:	A	B	C	D	___	60:	A	B	C	D	___
26:	A	B	C	D	___	61:	A	B	C	D	___
27:	A	B	C	D	___	62:	A	B	C	D	___
28:	A	B	C	D	___	63:	A	B	C	D	___
29:	A	B	C	D	___	64:	A	B	C	D	___
30:	A	B	C	D	___	65:	A	B	C	D	___
31:	A	B	C	D	___	66:	A	B	C	D	___
32:	A	B	C	D	___	67:	A	B	C	D	___
33:	A	B	C	D	___	68:	A	B	C	D	___
34:	A	B	C	D	___	69:	A	B	C	D	___
35:	A	B	C	D	___	70:	A	B	C	D	___

This page is intentionally left blank

1. The compound below is named:

A. pentanone

B. pentanal

C. butanaldehyde

D. pentaketone

2. Which of the compounds listed below is linear?

A. 1,3,5-heptatriene

B. acetylene

C. 2-butyne

D. dichloromethane

3. How many structural isomers of $C_4H_8Cl_2$ exhibit optical activity?

A. 0 B. 1 C. 2 D. 3

4. Infrared spectroscopy provides a scientist with information about:

A. molecular weight

B. distribution of protons

C. functional group

D. conjugation

5. Which of the following factors usually increase(s) the solubility of a molecule in a given solvent?

I. higher temperature
II. similar polarities
III. greater molecular weight of the molecule

A. I only

B. II only

C. III only

D. I and II only

6. What is the likely mechanism for the reaction between 1-bromobutane and sodium cyanide?

A. E_1 B. E_2 C. S_N1 D. S_N2

7. Which of the following reactions will NOT occur?

Reaction 1: butene + NBS $\rightarrow$ 2-bromobutane

Reaction 2: 2-methylbutane + 8 O_2 + heat $\rightarrow$ 5 CO_2 + 6 H_2O

Reaction 3: 2-methylbutane + Br_2 + hv $\rightarrow$ 2-bromo-2-methylbutane

Reaction 4: 2-methyl-2-butene + Br_2 + CCl_4 $\rightarrow$ 2,3-dibromo-2-methylbutane (+ enantiomer)

A. Reaction 1

B. Reaction 2

C. Reaction 3

D. Reaction 4

8. For isomers with the formula $C_{10}H_{16}$, which of the following structural features are NOT possible?

A. 2 rings and 1 double bond

B. 2 double bonds and 1 ring

C. 2 triple bonds

D. 1 ring and 1 triple bond

9. Which of the following molecules reacts the slowest in electrophilic nitration?

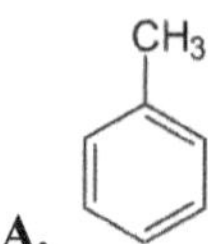

A. Toluene

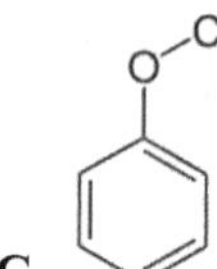

C. Anisole (methoxybenzene)

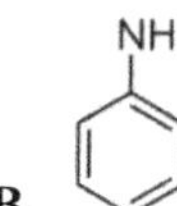

B. Aniline

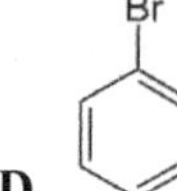

D. Bromobenzene

10. Which is a product of the oxidation of $CH_3–CH_2–CH_2–O–H$?

A. $CH_3–CH_2–CH_3$

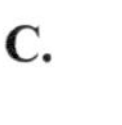

C.

B.

D.

11. Which of the following functional groups represents a ketone?

A.

C.

B.

D.

12. When a small amount of hexanoic acid [$CH_3(CH_2)_4CO_2H$, $pK_a \approx 4.8$] is added to a separatory funnel which contains the organic solvent diethyl ether and H_2O with a pH of 11.0, it is mainly in the [ ] phase as [ ]:

A. water; $CH_3(CH_2)_4CO_2H$

B. ether; $CH_3(CH_2)_4CO_2H$

C. water; $CH_3(CH_2)_4CO_2^-$

D. ether; $CH_3(CH_2)_4CO_2^-$

13. The reaction of benzoic acid with thionyl chloride followed, by treatment with ammonia, yields which of the following compounds?

A. *p*-chlorobenzamide

B. *m*-chlorobenzamide

C. benzamide

D. *p*-aminobenzaldehyde

14. The compound trimethylamine is a(n) [   ] and has the formula [   ]:

A. base, $(CH_3)_3N$

B. acid, $(CH_3)_2NH$

C. base, $(CH_3)_2NH$

D. acid, $(CH_3)_3N$

15. Provide the common name of the compound:

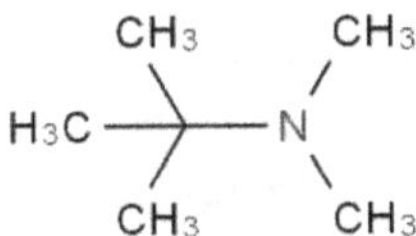

A. neobutyldimethylamine

B. *sec*-butyldimethylamine

C. *tert*-butyldimethylamine

D. isobutyldimethylamine

16. How many single and double bonds are in the benzene molecule (not including the C–H bonds)?

A. 6 single, 3 double

B. 5 single, 2 double

C. 6 single, 0 double

D. 5 single, 0 double

17. How many chiral carbon atoms are in this structure?

$$CH_2-CH-CH-CHCH_3$$
$$\ \ OH\ \ \ OH\ \ \ Br\ \ \ Cl$$

A. 6   **B.** 5   **C.** 3   **D.** 4

18. In the mass spectrum of 3,3-dimethyl-2-butanone, the base peak occurs at *m/z*:

A. 43   **B.** 58   **C.** 84   **D.** 85

19. Which of the three compounds is soluble in aqueous sodium bicarbonate?

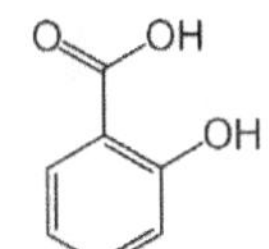

salicylic acid

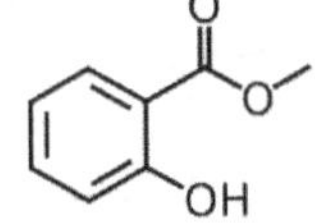

methyl salicylate

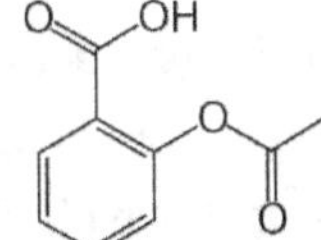

acetylsalicylic acid

    I. salicylic acid   II. methyl salicylate   III. acetylsalicylic acid

A. I only

B. II only

C. I and II only

D. I and III only

20. Which of the following alkyl halogens reacts the fastest with NaOH?

A. *t*-butyl bromide

B. *t*-butyl iodide

C. *t*-butyl fluoride

D. *t*-butyl chloride

21. Give the possibilities in the structure for a compound with a formula of C_6H_{10}:

    A. no rings; no double bonds; no triple bonds

    B. one double bond; or one ring

    C. two rings; two double bonds; one double bond and one ring; or one triple bond

    D. three rings; three double bonds; two double bonds and one ring; one ring and two double bonds; one triple bond and one ring; or one double bond and one ring

22. What is the product from the reaction of one mole of acetylene and one mole of hydrogen gas using platinum catalyst?

    A. propane   **C.** ethane

    B. propene   **D.** ethene

23. In electrophilic aromatic substitution, the aromatic ring acts as a(n):

    A. leaving group   **C.** nucleophile

    B. dienophile   **D.** electrophile

24. Which of the following has the highest boiling point?

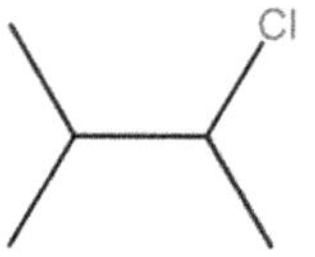

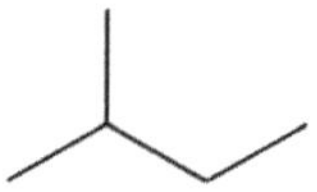

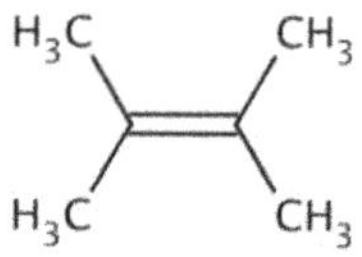

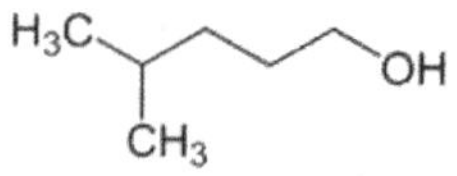

25. All of the following statements about oxidation of carbonyls are true, EXCEPT:

    A. oxidation of aldehydes produces carboxylic acids

    B. ketones do not react with mild oxidizing agents

    C. Tollens' test involves the oxidation of Ag^+

    D. Benedict's test involves the reduction of Cu^{2+}

26. Which of the following compounds acts as an acid?

    A. CH_3COCH_3 **C.** C_2H_5OH

    B. $(CH_3)_2NH$ **D.** C_2H_5COOH

27. The compound capsaicin below has which functional groups?

A. aldehyde and amine

B. carboxylic acid and amine

C. nitro and amine

D. amide and ether

28. Which compound is an example of an amine salt?

A. sulfanilamide

B. thioacetamide

C. dimethylammonium bromide

D. histamine

29. What is the correct IUPAC name for the following compound?

A. 2-oxocyclohex-3-ene-1-carboxylic acid

B. 5-formylcyclohex-2-enone-oic acid

C. 2-formylcyclohex-5-enone

D. 3-oxocyclohex-4-enoic acid

30. Determine the number of *pi* bonds in CH_3CN:

A. 0   **B.** 1   **C.** 2   **D.** 3

31. Which of the following is a true statement?

A. A mixture of achiral compounds is optically inactive

B. All molecules that possess a single chirality center of the *S* configuration are levorotatory

C. All achiral molecules are *meso*

D. All chiral molecules possess a plane of symmetry

32. The protons marked H_a and H_b in the molecule below are:

A. diastereotopic

B. heterotopic

C. homotopic

D. enantiotopic

33. Acetate can react with a tertiary alkyl chloride to form an ester. The reaction occurs more rapidly in water than in dimethylsulfoxide (DMSO) because water stabilizes the:

A. intermediate racemates

B. configuration inversion

C. carbocation intermediate

D. acetate

34. Which of the following is/are the most stable diene?

A. 

B. 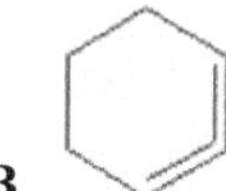

C. 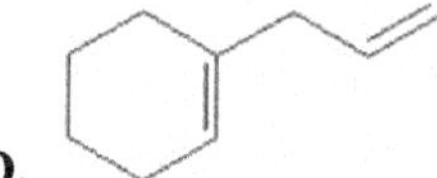

D.

35. Which of the species below is less basic than an acetylide?

I. CH_3Li II. CH_3MgBr III. CH_3ONa

A. I only

B. II only

C. III only

D. I and II only

36. What is the effect of –F substituents on electrophilic aromatic substitution?

A. *meta*-directing with activation

B. *meta*-directing with deactivation

C. *ortho- / para*-directing with activation

D. *ortho- / para*-directing with deactivation

37. What compound is formed by the oxidation of 2-hexanol?

A. hexanal

B. hexanoic acid

C. 2-hexanone

D. 2-hexene

38. Which of the following classes of compounds has a carbonyl group?

A. phenol

B. ether

C. amine

D. none of the above

39. All of the statements concerning citric acid are true, EXCEPT it:

A. is produced only by plants

B. is used in many consumer products

C. is highly soluble in water

D. contains three carboxylic acid groups because its carbon skeleton is branched

40. What is the name of the product formed by the reaction of propanoic acid with ethanol?

A. ethyl propanoate
B. pentanal
C. pentyl ester
D. ethyl propyl ketone

41. Which statement about the differences between an amine and an amide is NOT correct?

A. Amides act as proton acceptors, while amines do not
B. Amines are basic, and amides are neutral
C. Amines form ammonium salts when treated with acid; amides do not
D. The lone pair of electrons on amides is held more tightly than the lone pair on amines

42. What is the IUPAC name of the compound shown?

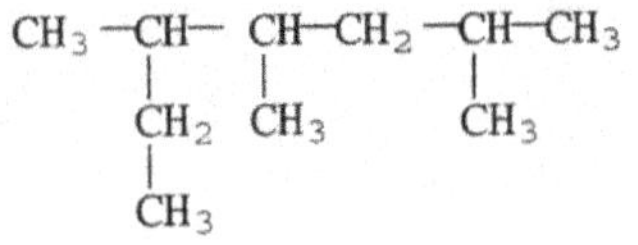

A. 3,4,6-trimethylheptane
B. 2,4,5-trimethylheptane
C. 3,5-dimethyl-2-ethylhexane
D. 2-ethyl-3,5-dimethylhexane

43. Give the hybridization, shape, and bond angle for carbon in ethene:

A. sp^3, tetrahedral, 120°
B. sp^3, tetrahedral, 109.5°
C. sp^2, trigonal planar, 120°
D. sp^2, trigonal planar, 109.5°

44. Which of the following best describes the geometry about the carbon-carbon double bond in this alkene?

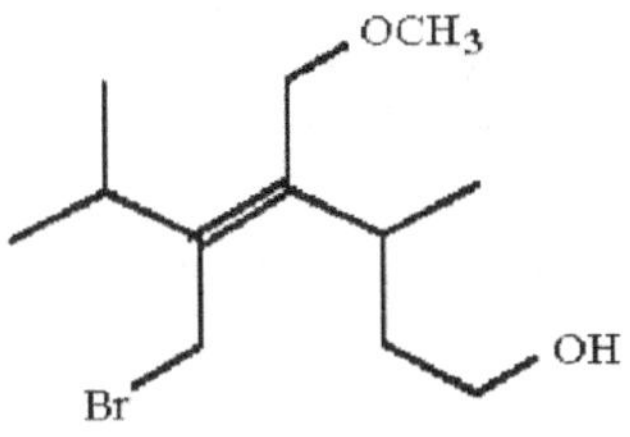

A. *E*
B. *Z*
C. *cis*
D. *R*

45. Identify the relationship between the compounds:

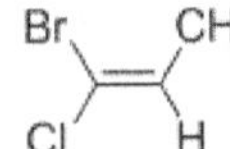

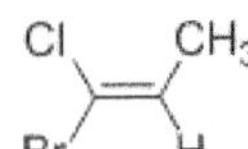

A. constitutional isomers
B. configurational isomers
C. identical
D. conformational isomers

46. The mass spectrum of alcohols often fails to exhibit detectable M peaks instead of showing relatively large [ ] peaks.

A. M–18 **B.** M+2 **C.** M–16 **D.** M–17

47. Which of the following pairs of alkenes has the same physical properties?

A. 2-butene and isobutene **C.** 1-butene and 2-butene

B. 1-butene and isobutene **D.** None of the above

48. Which of the following reactions is most likely to proceed by an S_N2 mechanism?

A. *t*-butyl iodide with ethanol **C.** 2-bromo-2-methyl pentane with HCl

B. 2-bromo-3-methyl pentane with methanol **D.** 1-bromopropane with NaOH

49. Which reactant converts propene to 1,2-dichloropropane?

A. HCl **B.** NaCl **C.** H_2 **D.** Cl_2

50. What are the two products from the complete combustion of an alkyne?

A. CO_2 and H_2O **C.** CO and H_2O

B. CO_2 and H_2 **D.** CO and H_2

51. In the molecular orbital representation of benzene, how many π molecular orbitals are present?

A. 1 **B.** 2 **C.** 4 **D.** 6

52. What is the major product of this reaction?

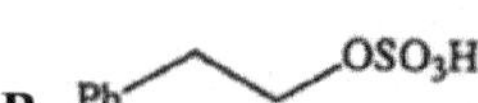

53. The reagent(s) that convert(s) a carbonyl group of a ketone into a methylene group is:

A. Na, NH_3, CH_3CH_2OH **C.** $NaBH_4$, CH_3CH_2OH

B. $LiAlH_4$ **D.** Zn(Hg), conc. HCl

54. The reaction of a carboxylic acid with a base such as sodium hydroxide, NaOH, gives:

A. alcohol

B. ester

C. alkoxide salt

D. carboxylate salt

55. What are the products of this reaction?

$$CH_3CH_2CONHCH_2CH_3 + NaOH \rightarrow ?$$

A. $CH_3CH_2COO^- Na^+ + CH_3CH_2NH_2$

B. $CH_3CH_2CH_2OH + CH_3CH_2NH_3^- Na^+$

C. $CH_3CH_2COO^- Na^+ + CH_3CH_2NH_3^+ Cl^-$

D. $CH_3CH_2COOH + CH_3CH_2NH_2$

56. Which of the following does NOT contain a polar carbonyl group?

A. an amine

B. a carboxylic acid

C. an ester

D. a ketone

57. What is the systematic name for the following compound?

A. 3-methyl-2-pentanol

B. 4-methyl-3-pentanol

C. 2-methyl-3-pentenol

D. 2-methyl-3-pentanol

58. In *trans*-hept-4-en-2-yne the shortest carbon-carbon bond is between carbons:

A. 1 and 2

B. 2 and 3

C. 3 and 4

D. 4 and 5

59. Which of the following carbons in the molecule below are chiral?

A. carbons 1 and 5

B. carbons 3 and 4

C. carbons 2, 3 and 4

D. all carbons are chiral

60. In ^1H NMR, protons on the α-carbon of amines typically resonate between:

A. 0.5 and 1.0 ppm

B. 2.0 and 3.0 ppm

C. 3.0 and 4.0 ppm

D. 6.0 and 7.0 ppm

61. Which of the following molecules can rotate freely around its carbon-carbon bond?

A. acetylene

B. cyclopropane

C. ethane

D. ethylene

62. Using Zaitsev's rule, choose the most stable alkene among the following:

A. 1-methylcyclohexene

B. 3-methylcyclohexene

C. 4-methylcyclohexene

D. 3,4-dimethylcyclohexene

63. The compound propyne consists of how many carbon atoms and how many hydrogen atoms?

A. 3C, 2H

B. 3C, 6H

C. 3C, 4H

D. 2C, 2H

64. Which of the following structures is aromatic?

A. 

C. 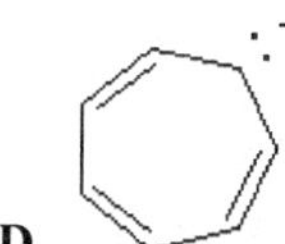

B. 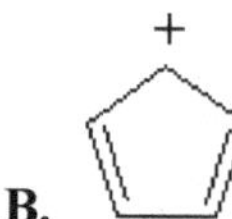

D.

65. Which of the following reagents is best to convert methyl alcohol to methyl chloride?

A. Cl^-

B. $SOCl_2$

C. Cl_2/CCl_4

D. $Cl_2/h\nu$

66. The reaction of ethylmagnesium bromide with which of the following compounds yields secondary alcohol after quenching with aqueous acid?

A. $(CH_3)_2CO$

B. ethylene oxide

C. H_2CO

D. CH_3CHO

67. What are the products from the reaction of ethanoic acid and methanol with sulfuric acid catalyst?

$$CH_3COOH + CH_3OH + H_2SO_4 \rightarrow ?$$

A. $CH_3CH_2COOH + H_2O$

B. $CH_3COOCH_3 + H_2O$

C. $CH_3COCH_3 + H_2O$

D. $CH_3CH_2CHO + H_2O$

68. Which product is formed during a reaction of acetic acid (CH_3COOH) with ammonia (NH_3)?

    A. ethylammonium hydroxide   **C.** acetamide

    B. amino acetate   **D.** ammonium acetate

69. When comparing amine compounds of different classes but similar molar masses, which type most likely has the highest boiling point?

    A. primary amines   **C.** tertiary amines

    B. secondary amines   **D.** quaternary ammonium salts

70. Which condensed structural formula is the isopropyl group?

    A.   **C.**

    B.   **D.**

Use the answer key to check your answers. Review the explanations in detail, focusing on questions you didn't answer correctly. Note the topic of those questions. Complete this BEFORE taking the next Diagnostic Test.

Notes for active learning

Diagnostic Test 6

This Diagnostic Test is designed to assess your proficiency on each topic and NOT to mimic the test. Use your test results and identify areas of strength and weakness to adjust your study plan and enhance your knowledge.

#	Answer:				Review	#	Answer:				Review
1:	A	B	C	D	___	36:	A	B	C	D	___
2:	A	B	C	D	___	37:	A	B	C	D	___
3:	A	B	C	D	___	38:	A	B	C	D	___
4:	A	B	C	D	___	39:	A	B	C	D	___
5:	A	B	C	D	___	40:	A	B	C	D	___
6:	A	B	C	D	___	41:	A	B	C	D	___
7:	A	B	C	D	___	42:	A	B	C	D	___
8:	A	B	C	D	___	43:	A	B	C	D	___
9:	A	B	C	D	___	44:	A	B	C	D	___
10:	A	B	C	D	___	45:	A	B	C	D	___
11:	A	B	C	D	___	46:	A	B	C	D	___
12:	A	B	C	D	___	47:	A	B	C	D	___
13:	A	B	C	D	___	48:	A	B	C	D	___
14:	A	B	C	D	___	49:	A	B	C	D	___
15:	A	B	C	D	___	50:	A	B	C	D	___
16:	A	B	C	D	___	51:	A	B	C	D	___
17:	A	B	C	D	___	52:	A	B	C	D	___
18:	A	B	C	D	___	53:	A	B	C	D	___
19:	A	B	C	D	___	54:	A	B	C	D	___
20:	A	B	C	D	___	55:	A	B	C	D	___
21:	A	B	C	D	___	56:	A	B	C	D	___
22:	A	B	C	D	___	57:	A	B	C	D	___
23:	A	B	C	D	___	58:	A	B	C	D	___
24:	A	B	C	D	___	59:	A	B	C	D	___
25:	A	B	C	D	___	60:	A	B	C	D	___
26:	A	B	C	D	___	61:	A	B	C	D	___
27:	A	B	C	D	___	62:	A	B	C	D	___
28:	A	B	C	D	___	63:	A	B	C	D	___
29:	A	B	C	D	___	64:	A	B	C	D	___
30:	A	B	C	D	___	65:	A	B	C	D	___
31:	A	B	C	D	___	66:	A	B	C	D	___
32:	A	B	C	D	___	67:	A	B	C	D	___
33:	A	B	C	D	___	68:	A	B	C	D	___
34:	A	B	C	D	___	69:	A	B	C	D	___
35:	A	B	C	D	___	70:	A	B	C	D	___

This page is intentionally left blank

1. A compound with the molecular formula C_3H_6 is:

  A. butane

  B. butyne

  C. 2-methylpropane

  D. cyclopropane

2. Which of the following is the least stable carbocation? (Use Ph = phenyl group)

  A. PhH_2C^+

  B. $CH_3CH_2CH_2^+$

  C. Ph_3C^+

  D. Ph_2HC^+

3. If two of the hydrogen atoms in ethylene, $H_2C=CH_2$, are replaced by two chlorine atoms to form dichloroethylene, how many different dichloroethylene isomers are there?

  A. 3   **B.** 4   **C.** 1   **D.** 2

4. A compound with nine carbon atoms produces a single NMR signal. Which is a structural formula for the compound?

  A. $(CH_3)_3CCCl_2C(CH_3)_3$

  B. $(CH_3)_2CHCH_2CH_2CH(CH_3)CH_2CH_3$

  C. $(CH_3)_2CHCH_2(CH_2)_4CH_3$

  D. $CH_3(CH_2)_7CH_3$

5. In a fractionating column, the crude oil vapors pass from a pipe into the column. Tar and lubricating stock are the first components to be pulled off at the bottom. Natural gas is the first fraction collected from the remaining hydrocarbons, followed by gasoline and then kerosene. From this information, which has a higher boiling point, gasoline, or kerosene?

  A. Fractional distillation components are pulled off based on molecular weight, so it is not possible to determine which has the higher boiling point from the information given

  B. their boiling points are the same, but kerosene has the greatest density

  C. Kerosene has a higher boiling point

  D. Gasoline has a higher boiling point

6. If propane was reacted with Cl_2 in the presence of UV light, what products form, and what are their approximate percentages?

  A. Propyl chloride yields 100%

  B. Propyl chloride yields 42%, and isopropyl chloride yields 58%

  C. Propyl chloride yields 75%, and isopropyl chloride yields 25%

  D. Propyl chloride yields 90%, and isopropyl chloride yields 10%

7. Which of the following reactions does NOT proceed through a bromonium ion intermediate?

A. $CH_3CH=CH_2 + Br_2 + H_2O \rightarrow CH_3CH_2OHCH_2Br + HBr$

B. $CH_2=CHCH_2CH_2CH=CH_2 + Br_2 \rightarrow CH_2BrCHBrCH_2CH_2CHBrCH_2Br$

C. $CH_3CH_2CH=CH_2 + Br_2 \rightarrow CH_3CH_2CHBrCH_2Br$

D. $CH_3CH_2CH=CH_2 + HBr \rightarrow CH_3CH_2CHBrCH_3$

8. What is the general molecular formula for the alkyne class of compounds?

I. C_nH_{2n+2} II. C_nH_{2n} III. C_nH_{2n-2}

A. I only

B. II only

C. III only

D. II and III only

9. What is the effect of each of –Cl substituents on electrophilic aromatic substitution?

A. *meta*-directing with deactivation

B. *meta*-directing with activation

C. *ortho-/para*-directing with deactivation

D. *ortho*-directing with activation

10. Which of the following alcohols has the lowest boiling point?

A. hexanol

B. 2-methyl-1-propanol

C. propanol

D. ethanol

11. Which of the following pairs have the most similar chemical properties?

A. alkanes and carboxylic acids

B. alkenes and aromatics

C. amines and esters

D. ketones and aldehydes

12. Which of the following compounds is the strongest acid?

A. $HOOCCH_2F$

B. $HOOCCH_3$

C. $HOOCCH_2Br$

D. $HOOCCH_2OCH_3$

13. Which of the following products might be formed if benzoyl chloride was treated with excess CH_3CH_2MgBr?

A. $C_6H_5CH_2CHO$

B. $C_6H_5C(CH_2CH_3)OHCH_2CH_3$

C. C_6H_5COOH

D. $C_6H_5COOCH_2CH_3$

---

14. Why is an amine salt more soluble in water than the corresponding free amine?

    A. The negative charge on the nitrogen atom increases water solubility

    B. It has a higher molecular weight than the corresponding amine

    C. It is ionic and therefore more soluble than covalent compounds with the same structure

    D. All amines are insoluble in water

15. What is the correct IUPAC name for the following compound?

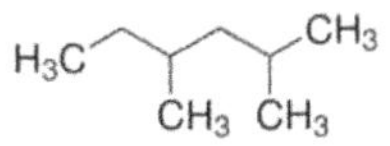

    A. 2-ethyl-4-methylhexane   **C.** 4-ethyl-2-methylhexane

    B. 2,4-dimethylhexane   **D.** 3,5-dimethylheptane

16. The energy of a sp^3 hybridized orbital for a carbon atom is:

    A. lower in energy than both the $2s$ and the $2p$ atomic orbitals

    B. higher in energy than both the $2s$ and the $2p$ atomic orbitals

    C. higher in energy than the $2p$ atomic orbital but lower in energy than the $2s$ atomic orbital

    D. higher in energy than the $2s$ atomic orbital but lower in energy than the $2p$ atomic orbital

17. The enantiomer of the compound below is:

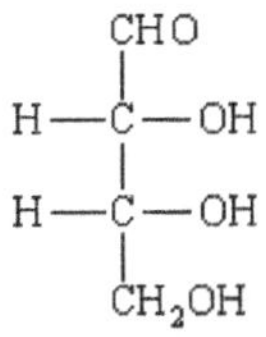

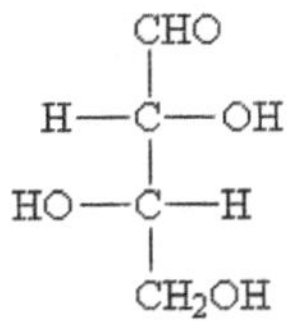

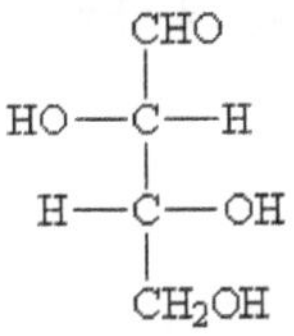

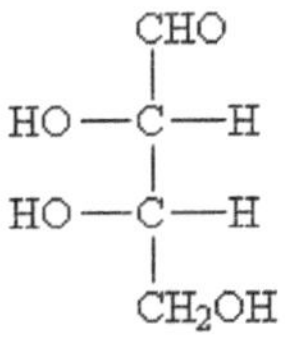

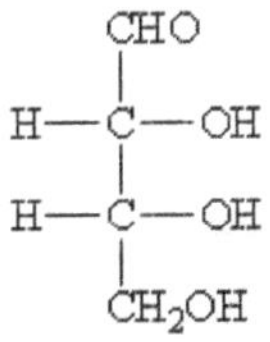

18. In mass spectrometry plots, the relative abundance is the unit along the y-axis in a mass spectrum. What are the units on the x-axis?

A. mass / charge (m/z)

B. mass (m)

C. molecular weight (amu)

D. frequency (v)

19. How does fractional distillation function to separate molecules?

A. It utilizes differences in melting points

B. It utilizes the fraction of carbon in the isomers

C. It utilizes the different weights of molecules

D. It utilizes the different boiling points of molecules

20. What is true about an S_N1 reaction?

I. A carbocation intermediate is formed

II. The rate-determining step is bimolecular

III. The mechanism has two steps

A. I only

B. II only

C. I and II only

D. I and III only

21. When reacted with HBr, *cis*-3-methyl-2-hexene most likely undergoes:

A. *anti*-Markovnikov *syn-* and *anti*-addition

B. *anti*-Markovnikov *syn*-addition

C. Markovnikov *syn-* and *anti*-addition

D. Markovnikov *syn*-addition

22. Among the following compounds, which acids are stronger than ammonia?

I. water   II. ethane   III. butyne   IV. but-2-yne

A. I and II only

B. II only

C. I and III only

D. II and III only

23. 1,3-cyclopentadiene reacts with sodium metal at low temperatures according to:

What is the best explanation for this observation?

A. Aromaticity stabilizes the anion

B. Sodium metal is highly selective for cycloalkenes

C. Aromaticity stabilizes the carbocation

D. The reactant is more unstable at reduced temperatures

24. The reaction of $(CH_3)_2CHCH_2OH$ with concentrated HBr using controlled heating yields:

A. $(CH_3)_2CHCH_2OBr$

B. $(CH_3)_2CHCH_4^+Br^-$

C. $CH_3CH_2CH_2Br$

D. $(CH_3)_2CHCH_2Br$

25. Which compound gives a positive Tollens test?

A. pentane

B. pentanal

C. 3-pentanone

D. 2-pentanone

26. Explain why caprylic acid $CH_3(CH_2)_6COOH$ dissolves in a 5% aqueous solution of sodium hydroxide, but caprylaldehyde, $CH_3(CH_2)_6CHO$, does not.

A. Caprylic acid reacts to form the water-soluble salt

B. Caprylaldehyde behaves as a reducing agent, which neutralizes the sodium hydroxide

C. Caprylaldehyde can form more hydrogen bonds to water than caprylic acid

D. With two oxygens, caprylic acid is about twice as polar as caprylaldehyde

27. The compound below has which functional groups?

A. lactone and alcohol

B. ether and alcohol

C. ester and hemiacetal

D. ester and acetal

28. Based on the properties of the attached functional group, which compound below interacts most strongly with water, thus making it the most soluble compound?

A. $CH_3–CH_2–I$

B. $CH_3–O–CH_3$

C. $CH_3–CH_2–NH_2$

D. $CH_3–CH_2–H$

29. Name the following structure according to IUPAC nomenclature:

A. 3-ethyl-3-hexene

B. 4-methylenehexane

C. 2-propyl-1-butene

D. 2-ethyl-1-pentene

30. Which of the following pairs are resonance structures?

A. 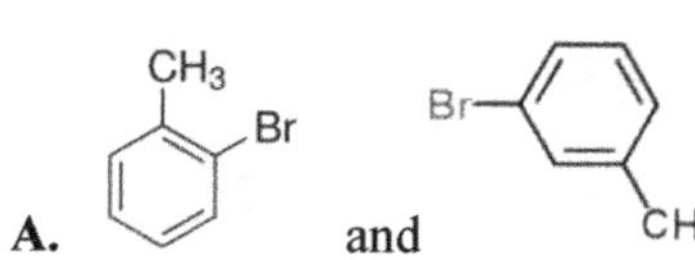 and

C. 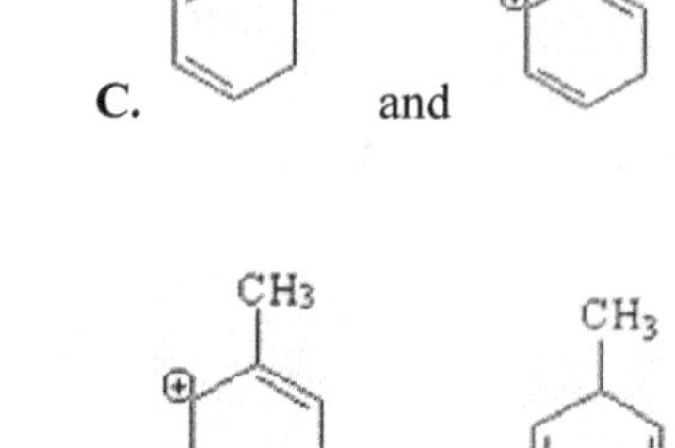 and

B. 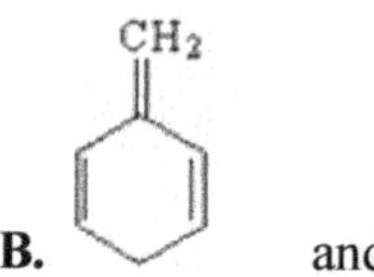 and

D. 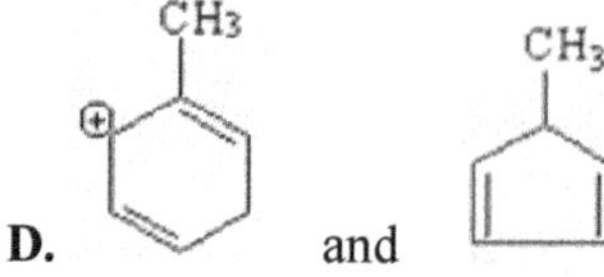 and

31. Which of the statements correctly describes an achiral molecule?

A. The molecule has an enantiomer

B. The molecule may be a *meso* form

C. The molecule has a non-superimposable mirror image

D. The molecule exhibits optical activity when it interacts with plane-polarized light

32. How many nuclear spin states are allowed for the ^{1}H nucleus?

A. 1   **B.** 2   **C.** 3   **D.** 4

33. The most stable conformational isomer of 1,2-dibromoethane is:

A. eclipsed, *anti*   **C.** staggered, *anti*

B. staggered, *gauche*   **D.** eclipsed, *gauche*

34. All of the following are examples of addition reactions of alkenes, EXCEPT:

A. ozonolysis   **C.** oxidation

B. hydration   **D.** bromination

35. In reducing alkynes using sodium in liquid ammonia, which of the species below is NOT an intermediate in the commonly accepted mechanism?

A. vinyl anion   **C.** anion

B. vinyl cation   **D.** vinyl radical

36. What is the major product of this electrophile aromatic substitution (EAS) reaction?

$+ HNO_3 / H_2SO_4 \rightarrow ?$

A.

C.

B.

D.

37. When propanol is subjected to PBr_3, the compounds undergo:

A. an S_N1 elimination reaction to form propene

B. oxidation to form an aldehyde

C. an S_N1 reaction to produce an alkyl halide

D. addition, elimination, and then substitution to form bromopropane

38. Consider the equilibrium of each of the carbonyl compounds with HCN to produce cyanohydrins. Which is the correct ranking of compounds in order of increasing K_{eq} for this equilibrium?

A. 2-methylcyclohexanone $<$ cyclohexanone $< CH_3CHO < H_2CO$

B. $CH_3CHO <$ 2-methylcyclohexanone $<$ cyclohexanone $< H_2CO$

C. cyclohexanone $<$ 2-methylcyclohexanone $< H_2CO < CH_3CHO$

D. cyclohexanone $<$ 2-methylcyclohexanone $< CH_3CHO < H_2CO$

39. All of the statements about carboxylic acids are true, EXCEPT:

A. they react with bases to form salts which are often more soluble than the original acid

B. they form hydrogen bonds with boiling points higher than expected based on molecular weight

C. when they behave as acids, the ⁻OH group is lost, leaving the CO⁻ ion

D. they undergo substitution reactions involving the ⁻OH group

40. Benzoyl chloride, PhC(O)Cl, reacts with water to form benzoic acid, $PhCO_2H$. In this addition–elimination reaction:

A. water acts as a nucleophile, benzoyl chloride acts as an electrophile, and chloride acts as a leaving group

B. water acts as an electrophile, benzoyl chloride acts as a nucleophile, and chloride acts as a leaving group

C. water and benzoyl chloride act as electrophiles

D. there are no nucleophiles or electrophiles

41. Assuming roughly equivalent molecular weights, which of the following has the highest boiling point?

A. alcohol

B. ether

C. tertiary amine

D. quaternary ammonium salt

42. What is the IUPAC name for the following structure?

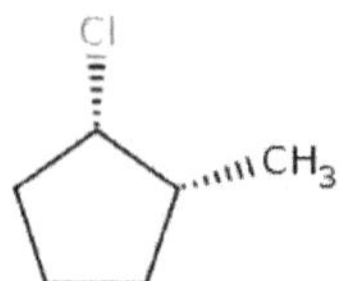

A. *cis*-methylcyclohexane

B. *cis*-1-chloro-2-methylcyclopentane

C. Z-chloro-methylcyclohexane

D. *cis*-2-chloro-2-methylcyclohexane

43. Due to electron delocalization, the carbon-oxygen bond in acetamide, CH_3CONH_2:

A. is longer than the carbon-oxygen bond of dimethyl ether, $(CH_3)_2O$

B. is longer than the carbon-oxygen bond of acetone, $(CH_3)_2CO$

C. is nonpolar

D. has more double bond character than the carbon-oxygen bond of acetone, $(CH_3)_2CO$

44. Which of the following statements does NOT correctly describe *cis*-1,2-dimethylcyclopentane?

A. Its diastereomer is *trans*-1,2-dimethylcyclopentane

B. It contains two asymmetric carbons

C. It is achiral

D. It has an enantiomer

45. What two atomic orbitals (or hybrid atomic orbitals) overlap to form the C=C π bond in ethylene?

A. C sp^2 + C p

B. C p + C p

C. C sp^3 + C sp^2

D. C sp^3 + C sp^3

46. How many different stereoisomers can the following compound have?

D-glucose

A. 2   **B.** 4   **C.** 8   **D.** 16

47. Ignoring geometric isomers, what is the IUPAC name for the following compound:

CH_3–CH=CH–CH_3, is:

A. but-2-yne   **C.** butene-3

B. but-2-ene   **D.** butene-2

48. Which of the following is an allylic cation?

A.   **C.**

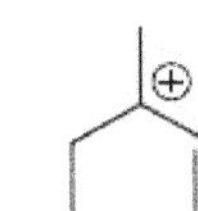

B.   **D.**

49. What is the relationship between the following molecules?

A. different molecules   **C.** identical

B. enantiomers   **D.** isomers

50. Free-radical chlorination of propane gives two isomeric monochlorides: 1-chloropropane and 2-chloropropane. How many NMR signals does each of these compounds display, respectively?

A. 3, 2   **B.** 3, 3   **C.** 2, 2   **D.** 2, 3

51. Which choices below list a sequence through which the three molecules shown (initially in a chloroform $CHCl_3$ solution) may be separated?

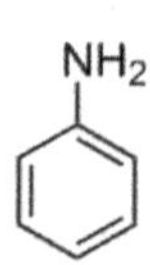

aniline

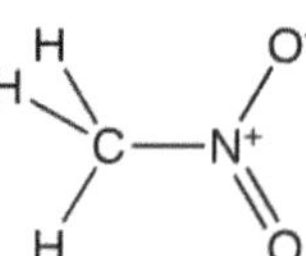

nitromethane

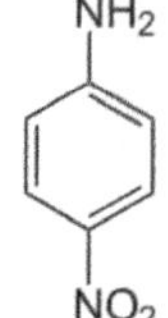

p-nitroaniline

A. Extraction with a strongly acidic aqueous solution → extraction with an even stronger acidic aqueous solution → distillation
B. Extraction with a strongly basic solution of NaOH → extraction with a weakly acidic aqueous solution of benzoic acid → distillation
C. Extraction with a weakly acidic aqueous solution of benzoic acid → extraction with a strongly acidic HCl aqueous solution → distillation
D. Extraction with a strongly acidic HCl aqueous solution → extraction with a weakly acidic aqueous solution → distillation

52. The rate of an S_N1 reaction depends on:

A. the concentration of the nucleophile and electrophile
B. neither the concentration of the nucleophile nor of the electrophile
C. the concentration of the nucleophile only
D. the concentration of the electrophile only

53. Which of the following correctly ranks the halides in order of increasing rate of addition to 3-hexene in a nonpolar aprotic solvent?

A. HI < HBr < HCl   **C.** HCl < HI < HBr
B. HBr < HCl < HI   **D.** HCl < HBr < HI

54. What is the term for a family of unsaturated hydrocarbon compounds with a triple bond?

A. alkynes   **C.** alkanes
B. arenes   **D.** alkenes

55. What is the effect of an ammonium substituent on electrophilic aromatic substitution?

A. *ortho/para*-directing with activation

B. *ortho/para*-directing with deactivation

C. *meta*-directing with activation

D. *meta*-directing with deactivation

56. Which formula is an alcohol?

A. R–O–H

B. 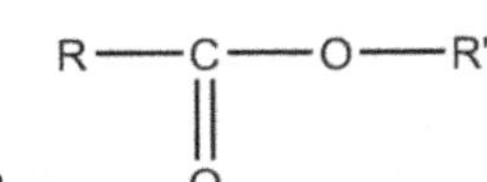

C. 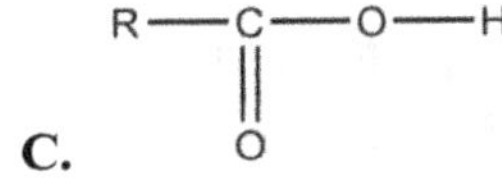

D. 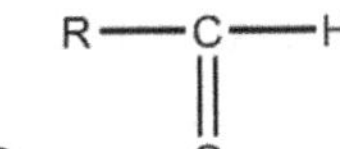

57. Which of the following reagents quantitatively converts an enolizable ketone to its enolate salt?

A. lithium hydroxide

B. lithium diisopropylamide

C. methyllithium

D. diethylamine

58. Which type of compound is shown below?

$$H-\overset{H}{\underset{H}{C}}-\overset{H}{\underset{H}{C}}-\overset{H}{\underset{H}{C}}-C\overset{O}{\underset{O-H}{}}$$

A. ester

B. carboxylic acid

C. ketone

D. aldehyde

59. Hydrolysis of the ester ethyl acetate produces:

A. butanal and ethanol

B. ethanal and acetic acid

C. ethanol and acetic acid

D. butanoic acid

60. Which of the following can be synthesized from an arenediazonium salt?

I. C_6H_5Br II. C_6H_5CN III. C_6H_5OH

A. I only

B. I and II only

C. I and III only

D. I, II and III

61. What is the IUPAC name for the following structure:

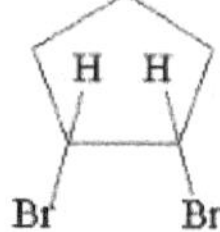

A. (*Z*)-3-ethyl-5-hydroxymethyl-3-penten-1-ynal
B. (*E*)-3-ethyl-5-hydroxymethyl-3-penten-1-ynal
C. (*Z*)-3-ethyl-2-hydroxymethyl-2-penten-4-ynal
D. (*E*)-3-ethyl-2-hydroxymethyl-2-penten-4-ynal

62. Which of the following best approximates the C–C–C bond angle of propene?

A. 90°
B. 109°

C. 120°
D. 150°

63. What is the relationship between the structures shown below?

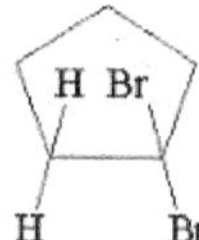

A. geometric isomers
B. conformational isomers

C. constitutional isomers
D. diastereomers

64. Which of the following transitions is usually observed in the UV spectra of ketones?

A. n to π^*
B. n to π

C. σ to n
D. σ to σ^*

65. Which of these statements correctly describes the separation of caffeine from heptanoic acid by extraction into dichloromethane (CH_2Cl_2) and water?

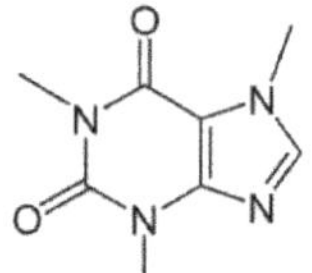

Caffeine

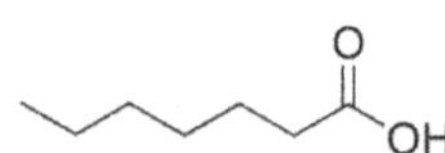

Heptanoic acid

A. In a basic solution, heptanoic acid is more soluble in CH_2Cl_2 than in H_2O
B. In an acidic solution, heptanoic acid is more soluble in H_2O than in CH_2Cl_2
C. In an acidic solution, caffeine is more soluble in CH_2Cl_2 than in H_2O
D. In a basic solution, caffeine is more soluble in CH_2Cl_2 than in H_2O

66. Which of the following properties is NOT characteristic of alkanes?

    A. They are tasteless and colorless

    B. They have strong hydrogen bonds

    C. Their melting points increase with the molecular weight

    D. They are generally less dense than water

67. The carbon–carbon single bond in 1,3-butadiene has a bond length that is shorter than a carbon–carbon single bond in an alkane. This is a result of:

    A. overlap of two sp^3 orbitals

    B. overlap of one sp^2 and one sp^3 orbital

    C. partial double-bond character due to the σ electrons

    D. overlap of two sp^2 orbitals

68. Which of the following molecular formulas correspond(s) to an alkyne?

    I. $C_{10}H_{18}$ II. $C_{10}H_{20}$ III. $C_{10}H_{22}$

    A. I only   **C.** III only

    B. II only   **D.** I and II only

69. Which of the following compounds undergoes Friedel-Crafts alkylation with $(CH_3)_3CCl$, $AlCl_3$ most rapidly?

    A. toluene   **C.** acetophenone

    B. iodobenzene   **D.** benzenesulfonic acid

70. The alcohol and carboxylic acid required to form propyl ethanoate are [   ] and [   ]:

    A. 1-propanol … ethanoic acid   **C.** ethanol … propionic acid

    B. propanol … propanoic acid   **D.** methanol … propionic acid

> Use the answer key to check your answers. Review the explanations in detail, focusing on questions you didn't answer correctly. Note the topic of those questions. Complete this BEFORE taking the next Diagnostic Test.

Notes for active learning

Diagnostic Test 7

This Diagnostic Test is designed to assess your proficiency on each topic and NOT to mimic the test. Use your test results and identify areas of strength and weakness to adjust your study plan and enhance your knowledge.

#	Answer:				Review	#	Answer:				Review
1:	A	B	C	D	___	36:	A	B	C	D	___
2:	A	B	C	D	___	37:	A	B	C	D	___
3:	A	B	C	D	___	38:	A	B	C	D	___
4:	A	B	C	D	___	39:	A	B	C	D	___
5:	A	B	C	D	___	40:	A	B	C	D	___
6:	A	B	C	D	___	41:	A	B	C	D	___
7:	A	B	C	D	___	42:	A	B	C	D	___
8:	A	B	C	D	___	43:	A	B	C	D	___
9:	A	B	C	D	___	44:	A	B	C	D	___
10:	A	B	C	D	___	45:	A	B	C	D	___
11:	A	B	C	D	___	46:	A	B	C	D	___
12:	A	B	C	D	___	47:	A	B	C	D	___
13:	A	B	C	D	___	48:	A	B	C	D	___
14:	A	B	C	D	___	49:	A	B	C	D	___
15:	A	B	C	D	___	50:	A	B	C	D	___
16:	A	B	C	D	___	51:	A	B	C	D	___
17:	A	B	C	D	___	52:	A	B	C	D	___
18:	A	B	C	D	___	53:	A	B	C	D	___
19:	A	B	C	D	___	54:	A	B	C	D	___
20:	A	B	C	D	___	55:	A	B	C	D	___
21:	A	B	C	D	___	56:	A	B	C	D	___
22:	A	B	C	D	___	57:	A	B	C	D	___
23:	A	B	C	D	___	58:	A	B	C	D	___
24:	A	B	C	D	___	59:	A	B	C	D	___
25:	A	B	C	D	___	60:	A	B	C	D	___
26:	A	B	C	D	___	61:	A	B	C	D	___
27:	A	B	C	D	___	62:	A	B	C	D	___
28:	A	B	C	D	___	63:	A	B	C	D	___
29:	A	B	C	D	___	64:	A	B	C	D	___
30:	A	B	C	D	___	65:	A	B	C	D	___
31:	A	B	C	D	___	66:	A	B	C	D	___
32:	A	B	C	D	___	67:	A	B	C	D	___
33:	A	B	C	D	___	68:	A	B	C	D	___
34:	A	B	C	D	___	69:	A	B	C	D	___
35:	A	B	C	D	___	70:	A	B	C	D	___

This page is intentionally left blank

1. The IUPAC name for the compound $H_2C=CH–CH=CH_2$ is:

A. 1,3-butadiene

B. butane-1,3

C. butene-2

D. 1,3-dibutene

2. Draw a structural formula for cyclohexane, a cyclic saturated hydrocarbon (C_6H_{12}). How many π bonds are in a cyclohexane molecule?

A. 3   **B.** 4   **C.** 0   **D.** 2

3. Which of the following compounds is an isomer of $CH_3CH_2CH_2CH_2OH$?

A. $CH_3CH_2CH_2OH$

B. $CH_3(OH)CHCH_3$

C. $CH_3CH_2CH_2CHO$

D. $CH_3CH_2(OH)CHCH_3$

4. A clear liquid is subjected to infrared spectroscopy and produces a spectrum with a prominent, broad peak at approximately 3000 cm^{-1} and a sharp peak at 1710 cm^{-1} and several smaller peaks between 1420 cm^{-1} and 940 cm^{-1}. This substance is most likely:

A. ketone

B. carboxylic acid

C. alcohol

D. aldehyde

5. Which of the following factors usually increase(s) the solubility of a molecule in a given solvent?

> I. higher temperature
> II. similar polarities
> III. lower density of the solvent

A. I only

B. II only

C. III only

D. I and II only

6. Which of the compounds listed below has the lowest boiling point?

A. 3-methylheptane

B. 2,4-dimethylhexane

C. octane

D. 2,2,4-trimethylpentane

7. Which of the following compounds is/are geometric isomers?

> I. Isobutene
> II. (*E*)-2-butene
> III. *cis*-2-butene
> IV. *trans*-2-butene

A. I and II only

B. II and III only

C. III and IV only

D. II, III and IV only

8. When 2,2-dibromobutane is heated to 200 °C in the presence of molten KOH, what is the major organic product?

   A. but-1-yne   **C.** 1-bromobut-1-yne

   B. but-2-yne   **D.** 1-bromobut-2-yne

9. Derivatives of the compound shown below are currently being examined for their effectiveness in treating drug addiction and metabolic syndrome. Which sequence ranks the aromatic rings of this compound in order of increasing reactivity (slowest to fastest reacting) in an electrophilic aromatic substitution reaction?

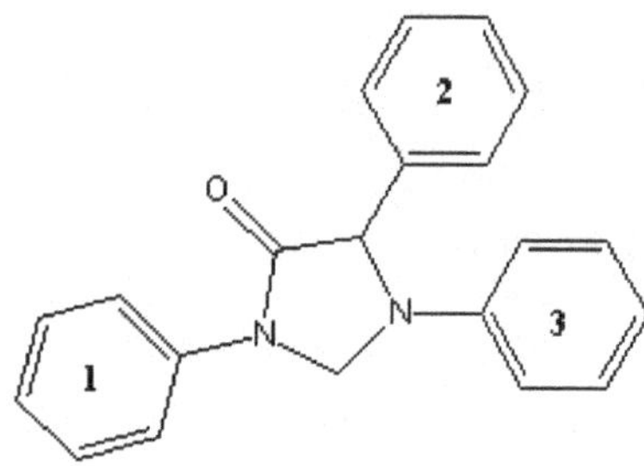

   A. 2 < 1 < 3   **C.** 3 < 2 < 1

   B. 2 < 3 < 1   **D.** 3 < 1 < 2

10. Which of the following has the highest boiling point?

   A. ethyl methyl ether   **C.** dimethyl ether

   B. dihexyl ether   **D.** diethyl ether

11. Which compound gives a positive indicator with the Tollens' reagent?

   A. $H-C(=O)-O-CH_2-CH_3$ **C.** $CH_3-C(=O)-CH_2-CH_3$

   B. $H-C(OH)-CH_2-CH_3$ **D.** $CH_3-C(=O)-H$

12. Which of the functional groups below contain(s) a hydroxyl group as a part of its/their structure?

   I. anhydride   II. carboxylic acid   III. ester

   A. I only   **C.** III only

   B. II only   **D.** I and II only

13. Which functional group(s) below indicate(s) the presence of two atoms connected by a triple bond?

    I. ester   II. nitrile   III. alkyne

A. I only   **C.** III only

B. II only   **D.** II and III only

14. The functional group C=O is in all the species below, EXCEPT:

A. amines   **C.** carboxylic acids

B. amides   **D.** aldehydes

15. What is the IUPAC name for salicylic acid shown below:

A. 2-hydroxybenzoic acid   **C.** 1-hydroxybenzoic acid

B. α-hydroxybenzoic acid   **D.** *meta*-hydroxybenzoic acid

16. The N–H bond in the ammonium ion, NH_4^+, is formed by the overlap of which orbitals?

A. sp^2–s **B.** sp^2–sp^2 **C.** sp^3–s **D.** sp^3–sp^3

17. If two hydrogen atoms in ethylene, $H_2C=CH_2$, are replaced by one chlorine atom and one fluorine atom to form chlorofluoroethene, C_2H_2ClF, how many different chlorofluoro-ethene isomers are there?

A. 4   **B.** 3   **C.** 2   **D.** 1

18. Which of the following laboratory techniques is used primarily as a compound identification procedure?

A. extraction   **C.** crystallization

B. NMR spectroscopy   **D.** distillation

19. The best method to separate each volatile isomer product from the reaction mixture is:

A. extraction   **C.** fractional distillation

B. crystallization   **D.** thin-layer chromatography

20. Which of the following is NOT a feature of S_N2 reactions?

A. single-step mechanism   **C.** bimolecular kinetics

B. pentacoordinate transition state   **D.** carbocation intermediate

21. Consider the following alcohol A.

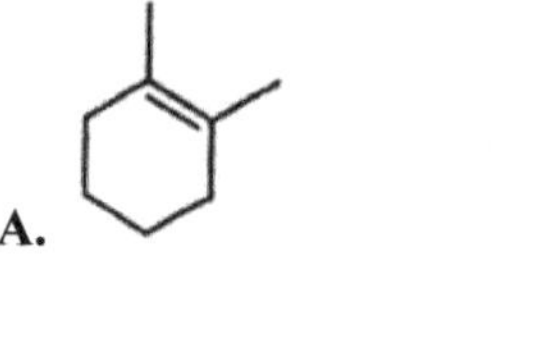

The major product resulting from the dehydration of A is:

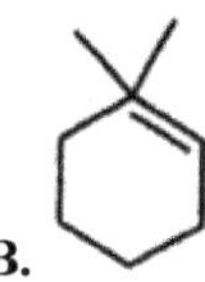

A.

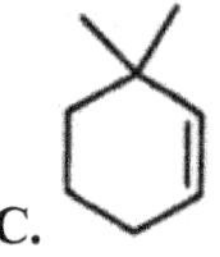

C.

B.

D.

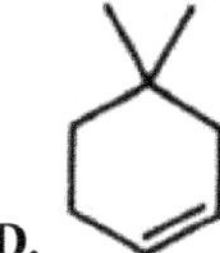

22. What class of organic product results when 1-heptyne is treated with a mixture of mercuric acetate [Hg(OAc)$_2$] in aqueous sulfuric acid (H$_2$SO$_4$), followed by sodium borohydride (NaBH$_4$)?

A. diol

B. ether

C. aldehyde

D. ketone

23. The major aromatic product of the following reaction is:

A. methyl ketone substitutes in the *ortho* / *para* position
B. methyl ketone substitutes in the *meta* position
C. methyl ketone replacing the bromine
D. formation of phenol

24. Based on the properties of the attached functional group, which compound interacts most strongly with water, thus making it the most soluble compound?

A. CH$_3$–CH$_2$–S–H

B. CH$_3$–CH$_2$–I

C. CH$_3$–CH$_2$–O–H

D. CH$_3$–CH$_2$–Cl

25. Reduction of aldehydes and ketones is a [　] reaction involving the [　] ion(s).

A. two-step; H^- and H^+

B. two-step; OH^- and H^+

C. one-step; H^-

D. one-step; H^+

26. Which acid is expected to have the lowest boiling point?

A. formic, HCO_2H

B. oxalic, $(CO_2H)_2$

C. acetic, CH_3CO_2H

D. benzoic, $C_6H_5CO_2H$

27. Which of these molecules is an ester?

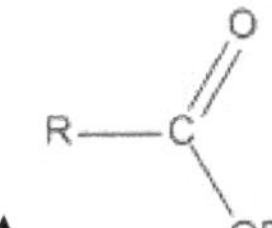

A.

C.

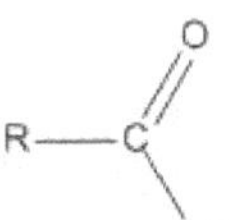

B.

D.

28. A quaternary ammonium salt does NOT undergo substitution reactions with alkyl halides because the nitrogen atom is not:

A. an electrophile

B. a nucleophile

C. negatively charged

D. saturated

29. What is the common name for the simplest ketone, propanone?

A. acetal

B. acetone

C. carbanone

D. formalin

30. Among the hydrogen halides, the strongest bond is in [　], and the longest bond is in [　]?

A. HI ... HI

B. HI ... HF

C. HF ... HI

D. HF ... HF

31. In the Fischer projection below, what are the configurations of the two asymmetric centers?

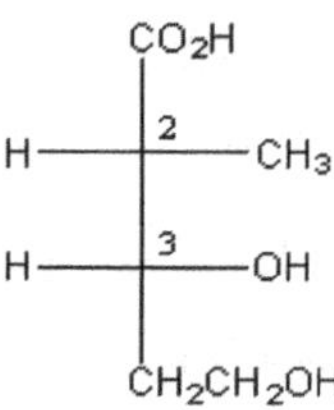

A. 2S, 3S

B. 2S, 3R

C. 2R, 3S

D. 2R, 3R

32. While the carbonyl stretching frequency for simple aldehydes, ketones, and carboxylic acids is about 1710 cm^{-1}, the carbonyl stretching frequency for esters is about:

A. 1660 cm^{-1}

B. 1700 cm^{-1}

C. 1735 cm^{-1}

D. 1800 cm^{-1}

33. Which conformer is at a local energy minimum on the potential energy diagram in the chair-chair interconversion of cyclohexane?

A. boat

B. twist-boat

C. half-chair

D. planar

34. What is the major product of the following reaction?

$+ Br_2 / H_2O \rightarrow$?

A.

B.

C.

D.

35. Which of the following describes the reaction below?

$H_3C \!\!\equiv\!\! CH_3 + H_2$, Pd, $CaCo_3$, quinolone, hexane $\rightarrow$

A. oxidation

B. reduction

C. substitution

D. catalytic hydration

36. In electrophilic aromatic substitution reactions, an extremely reactive electrophile is typically used because the aromatic ring is:

A. a poor electrophile

B. nonpolar

C. reactive

D. a poor nucleophile

37. What is the product of the following reaction?

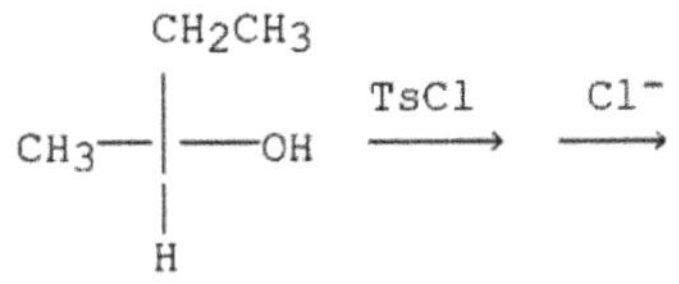

$$CH_3-\overset{\overset{\displaystyle CH_2CH_3}{|}}{\underset{\underset{\displaystyle H}{|}}{C}}-OH \xrightarrow{\text{TsCl}} \xrightarrow{\text{Cl}^-}$$

A. $TsO-\overset{\overset{\displaystyle CH_2CH_3}{|}}{\underset{\underset{\displaystyle H}{|}}{C}}-CH_3$

C. $CH_3-\overset{\overset{\displaystyle CH_2CH_3}{|}}{\underset{\underset{\displaystyle H}{|}}{C}}-OTs$

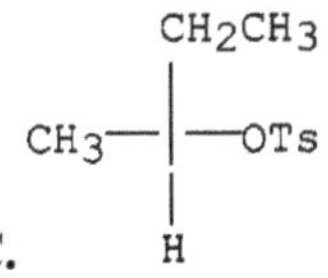

B. 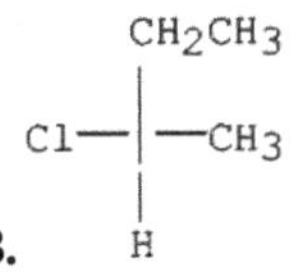 $Cl-\overset{\overset{\displaystyle CH_2CH_3}{|}}{\underset{\underset{\displaystyle H}{|}}{C}}-CH_3$

D. 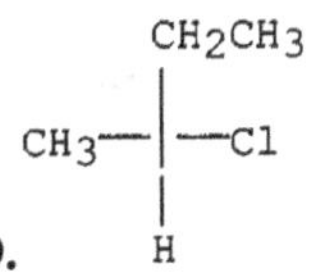 $CH_3-\overset{\overset{\displaystyle CH_2CH_3}{|}}{\underset{\underset{\displaystyle H}{|}}{C}}-Cl$

38. Which of the following is the best Michael acceptor?

A. 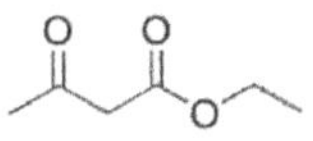 (diethyl malonate / ethyl 3-oxobutanoate structure)

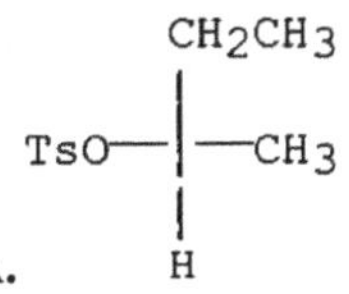

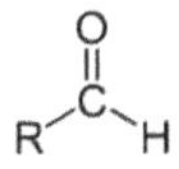

B. 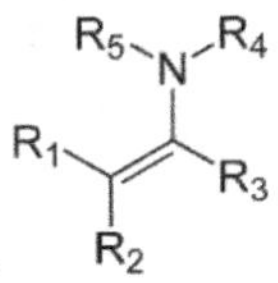 (enamine: $R_1R_2C=CR_3$ with NR_4R_5)

C. $R-\overset{\overset{\displaystyle O}{\|}}{C}-H$

D. 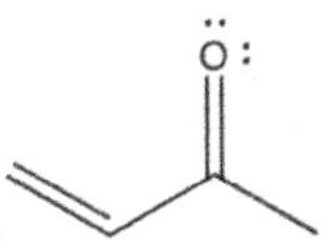 (methyl vinyl ketone structure)

39. Which of the following molecules is acidic?

I. 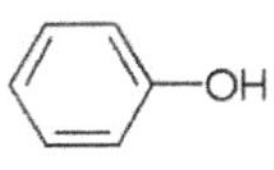 (phenol, C₆H₅–OH)

II. 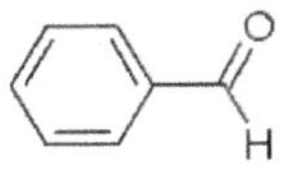 (benzaldehyde)

III. 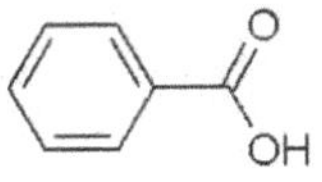 (benzoic acid)

A. I only

B. II only

C. I and II only

D. I and III only

40. The products of basic hydrolysis of an ester are:

A. acid + water

B. alcohol + water

C. carboxylate salt + alcohol

D. another ester + water

41. Amines are most similar in chemical structure and behavior to:

A. sodium hydroxide

B. a primary alcohol

C. the hydronium ion

D. ammonia

42. The name of the compound shown below is:

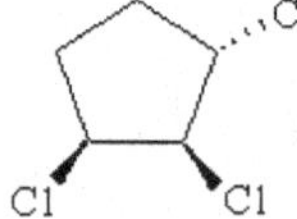

A. 3-ethyltoluene

B. 2-ethyltoluene

C. 1-ethyl-4-methylbenzene

D. 1-ethyl-2-methylbenzene

43. Which of the following is the most stable carbocation?

A.

B.

C.

D.

44. Which of the following terms best describes the pair of compounds shown?

A. same molecule

B. conformational isomers

C. enantiomers

D. diastereomers

45. Which type of compound is shown below?

$$H_3C \overset{\displaystyle O}{\underset{\displaystyle \|}{\diagup\!\!\diagdown}} CH_3$$

A. ester	**C.** ketone
B. carboxylic acid	**D.** aldehyde

46. Which carboxylic acid is used to prepare the ester shown?

$$CH_3-CH_2-O-C-CH_2-CH-CH_3$$

A. $(CH_3)_2CHCH_2COOH$	**C.** CH_3COOH
B. $(CH_3)_2CHCOOH$	**D.** CH_3CH_2COOH

47. What class of compound has the following general structure?

$$R-C-O-R'$$

A. ester	**C.** aldehyde
B. ketone	**D.** anhydride

48. Methyl bromide can generate the corresponding methylamine through alkyl halide ammonolysis. The nitrile reduction pathway generates the corresponding:

A. ethylmethylamide	**C.** reduced methylene group
B. ethylamine	**D.** dehalogenated methyl group

49. Give the IUPAC name for the following structure:

A. 1-chloro-4-methylcyclohexanol	**C.** 3-chloro-2-methylcyclohexanol
B. 5-chloro-2-methylcyclohexanol	**D.** 2-methyl-5-chlorocyclohexanol

50. What determines the polarity of a covalent bond?

A. The difference in the number of protons

B. The difference in the number of valence electrons

C. The difference in atomic size

D. The difference in electronegativity

51. What is the relationship between the following compounds?

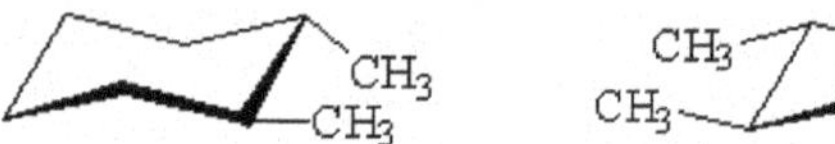

A. conformational isomers

C. enantiomers

B. diastereomers

D. constitutional isomers

52. What is the relative area of each peak in a quartet spin-spin splitting pattern?

A. 1:1:1:1

C. 1:2:1

B. 1:3:3:1

D. 1:2:2:1

53. Predict the most likely mechanism for the reaction shown below:

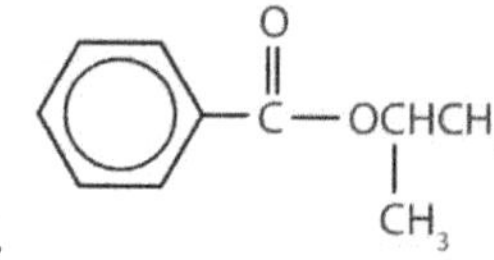 $+ NaOCH_3 / CH_3OH \rightarrow$?

A. E_2 **B.** E_1 **C.** S_N2 **D.** S_N1

54. What is the name of the major organic product of the following reaction?

$(CH_3)_2C=C(CH_3)_2 + H^+ / H_2O \rightarrow$?

A. 2,3-dimethyl-2-butanol

C. 3,3-dimethyl-1-butanol

B. 2,3-dimethyl-1-butanol

D. 3,3-dimethyl-2-butanol

55. What is the product from the reaction of one mole of acetylene and two moles of hydrogen gas using a platinum catalyst?

A. propene

C. ethene

B. propane

D. ethane

56. Which sequence correctly ranks the following aromatic rings in order of increasing the rate of reactivity in an electrophilic aromatic substitution reaction?

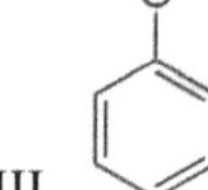

A. I < II < III

C. III < I < II

B. II < III < I

D. II < I < III

57. Which compound has the highest boiling point?

A. $CH_3CH_2CH_2CH_2OH$

B. $CH_3CH_2CH_2CH_3$

C. $CH_3CH_2CH_2CH_2CH_2OH$

D. $CH_3CH_2CH_2CH_2CH_3$

58. Which of the following carbonyl compounds may be synthesized from 1,3-dithiane?

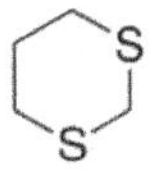

I. methyl vinyl ketone

II. 2-pentanone

III. 3,3-dimethyl-2-butanone

IV. 2-phenylethanal

A. I and IV only

B. II only

C. II and III only

D. II and IV only

59. When an amine reacts with a carboxylic acid at high temperature, the major product is a(n):

A. ether

B. thiol

C. amide

D. ester

60. What functional group is NOT present in the following structure for thyroxine (i.e., thyroid hormone)?

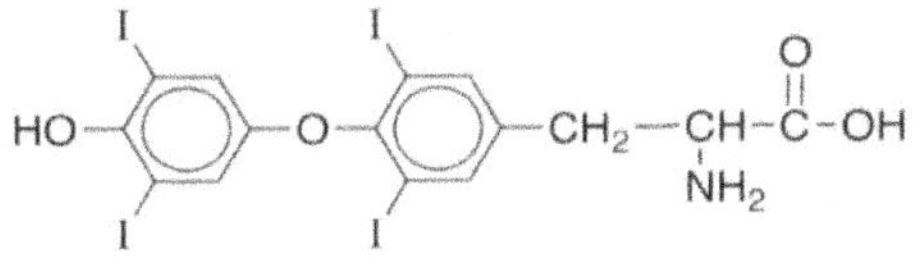

A. organic halide

B. ether

C. carboxylic acid

D. anhydride

61. Amines are classified by the:

A. number of carbons present in the molecule

B. number of carbons attached to the carbon bonded to the nitrogen

C. number of hydrogens attached to the nitrogen

D. number of alkyl groups attached to the nitrogen

62. What is the name of the following compound?

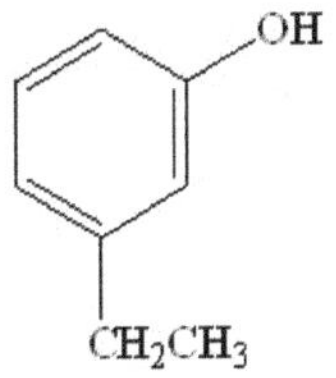

A. *p*-ethylphenol

B. *m*-ethylbenzene

C. *o*-ethylphenol

D. *m*-ethylphenol

63. What is the molecular geometry of an open-chain noncyclical hydrocarbon with the generic molecular formula C_nH_{2n-2}?

    A. trigonal pyramidal

    B. trigonal planar

    C. tetrahedral

    D. linear

64. Which of the following is NOT true of enantiomers?

    A. Enantiomers have the same chemical reactivity with non-chiral reagents

    B. Enantiomers have the same direction of specific rotation

    C. Enantiomers have the same melting point

    D. Enantiomers have the same boiling point

65. What products are formed upon the reaction of benzoic acid with sodium hydroxide, NaOH?

    A. Sodium bicarbonate and sodium benzoate

    B. Sodium bicarbonate and benzaldehyde

    C. Sodium benzoate and water

    D. Benzaldehyde and water

66. Which of the reactions is favorable, in the direction indicated, under common laboratory conditions?

A. methyl benzoate $\xrightarrow{\text{CH}_3\text{NH}_2}$ N-methylbenzamide

B. methyl benzoate $\xrightarrow{\text{CH}_3\text{COOH}}$ benzoic acetic anhydride

C. benzamide $\xrightarrow{\text{CH}_3\text{CH}_2\text{OH}}$ ethyl benzoate

D. benzoic acid $\xrightarrow{\text{CH}_3\text{COOH}}$ benzoic acetic anhydride

67. The following is an example of a:

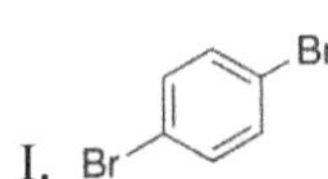

A. quaternary ammonium salt

B. tertiary ammonium

C. quaternary amide salt

D. tertiary amine

68. Which structure is *para*-dibromobenzene?

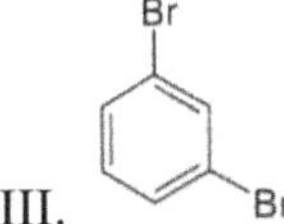

A. I only

B. II only

C. III only

D. I and II only

69. What chemical reaction was used by German chemist Friedrich Wohler to synthesize urea for the first time?

A. Heating ammonium cyanide

B. Combining the elements carbon, hydrogen, oxygen, and nitrogen

C. Evaporating urine

D. Heating ammonium cyanate

70. *Cis-trans* isomerism occurs when:

A. each carbon in an alkene double bond has two different substituent groups

B. the carbons in the *para* position of an aromatic ring have the same substituent groups

C. a branched alkane has a halogen added to two adjacent carbon atoms

D. an alkene is hydrated according to Markovnikov's Rule

> Use the answer key to check your answers. Review the explanations in detail, focusing on questions you didn't answer correctly. Note the topic of those questions. Complete this BEFORE taking the next Diagnostic Test.

Notes for active learning

Supplemental Questions

1. Name the structure:

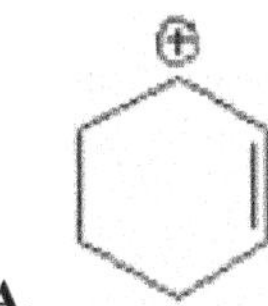

   A. *cis*-7-chloro-3-ethyl-4-methyl-3-heptene   **C.** 1-chloro-5-ethyl-4-methyl-3-heptene

   B. 1-chloro-3-pentenyl-2-pentene   **D.** 7-chloro-3-ethyl-4-methyl-3-heptene

2. Which of the following is a benzylic cation?

   A.

   C.

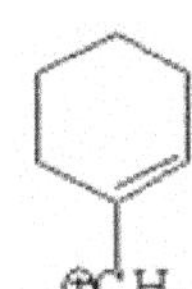

   B.

   D.

3. Butene, C_4H_8, is a hydrocarbon with one double bond. How many isomers are there of butene?

   A. two   **C.** four

   B. three   **D.** five

4. In the proton NMR, in what region of the spectrum does one typically observe hydrogens bound to the aromatic ring?

   A. 1.0-1.5 ppm   **C.** 4.5-5.5 ppm

   B. 2.0-3.0 ppm   **D.** 6.0-8.0 ppm

5. Which of the following factors usually increase(s) the solubility of a compound in a given solvent?

   I. higher temperature   III. higher molecular weight of the compound

   II. similar polarities   IV. lower density of the solvent

   A. I only   **C.** I, II and III only

   B. I and II only   **D.** I, II and IV only

6. A nucleophile is:

A. an oxidizing agent

B. electron deficient

C. a Lewis base

D. a Lewis acid

7. Both (*E*)- and (*Z*)-hex-3-ene can be subjected to a hydroboration-oxidation sequence. How are the products from these two reactions related?

A. The products of the two isomers are diastereomers

B. The products of the two isomers are constitutional isomers

C. The (*E*)- and (*Z*)-isomers generate the same products in the same amounts

D. The (*E*)- and (*Z*)-isomers generate the same products but in differing amounts

8. What is the major organic product that results when 3-heptyne is subjected to excess hydrogen and a platinum catalyst?

A. heptane

B. (*Z*)-3-heptene

C. (*Z*)-2-heptene

D. 2-heptyne

9. What is the major product of this reaction?

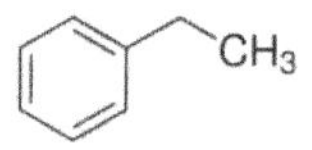

$$\text{C}_6\text{H}_5\text{CH}_2\text{CH}_3 + \text{C(CH}_3)_3\text{Br} + \text{FeBr} \rightarrow \text{?}$$

A.

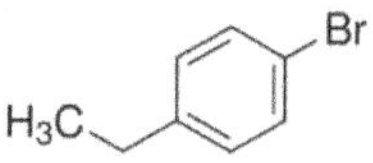

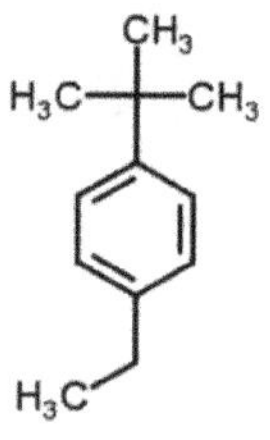

B.

C.

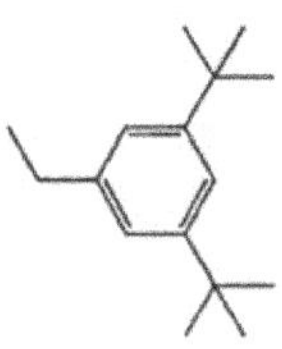

D.

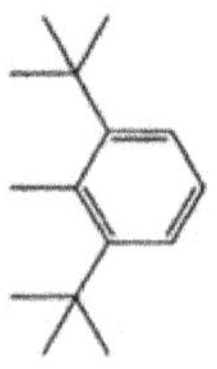

10. Which of the following alcohols has the highest boiling point?

A. 2-methyl-1-propanol

B. hexanol

C. ethanol

D. propanol

11. Oxidation of a ketone produces:

    A. a secondary alcohol   **C.** a carboxylic acid

    B. an aldehyde   **D.** no reaction

12. Which of the following functional groups does this organic compound contain?

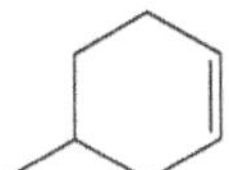

    A. amide   **C.** carboxylic acid

    B. amine   **D.** ester

13. The compound below is which type of compound?

    A. amide   **C.** amine

    B. amino acid   **D.** aldehyde

14. Which compound is a primary amine?

    A. *N, N*-dimethylethylamine   **C.** diethylamine

    B. isopropylamine   **D.** trimethylamine

15. What is the IUPAC name for the following compound?

    A. 1-methyl-4-cyclohexene   **C.** 4-methylcyclohexene

    B. 1-methyl-3-cyclohexene   **D.** 5-methylcyclohexene

16. Which orbitals overlap to create the H–C bond in CH_3^+?

    A. $s\text{–}p$ **C.** $sp^3\text{–}sp^3$

    B. $s\text{–}sp^2$ **D.** $sp^2\text{–}sp^3$

17. Which of the following compounds has an asymmetric center?

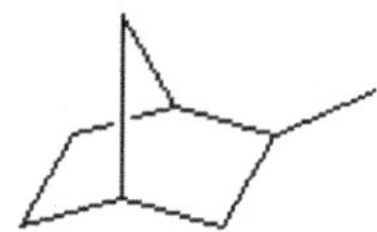

A.

C.

B.

D.

18. ^1H nuclei near electronegative atoms are [ ] relative to ^1H nuclei, which are not near them.

A. coupled

C. shielded

B. split

D. deshielded

19. Separation by thin-layer chromatography is based primarily on:

A. relative specific rotations (+/−) of the sample

B. comparative refractive indices (R_f) of the sample and the solvent

C. relative attraction of the sample towards the stationary and mobile phases

D. the molecular weight of the sample

20. Identify the number of tertiary carbons in the following structure:

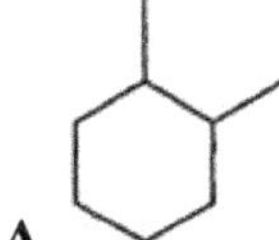

A. 4   B. 5   C. 2   D. 3

21. What is the major product of this reaction?

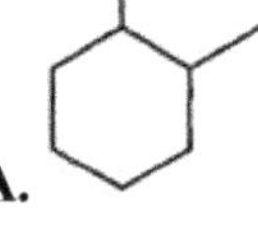

+ D_2 / Pt → ?

A.

C.

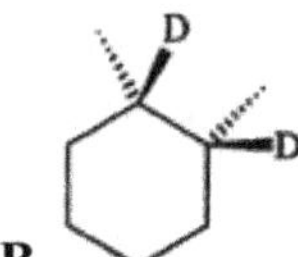

B.

D.

22. In the addition of hydrogen bromide to alkynes in the absence of peroxides, which of the following species is considered an intermediate?

A. carbene

B. vinyl radical

C. vinyl cation

D. vinyl anion

23. All of the following are common reactions of benzene, EXCEPT:

A. nitration

B. hydrogenation

C. chlorination

D. bromination

24. When phenol acts as an acid, a [   ] ion is produced.

A. phenoxide

B. benzol

C. phenyl

D. benzyl

25. (*S*)-2-methylbutanal [   ] upon sitting in an acidic or basic aqueous solution.

A. racemizes

B. esterifies

C. inverts completely to the *R* configuration

D. hydrolyzes

26. What is the major organic product of the reaction shown?

$$CH_3(CH_2)_3COOH + CH_3C(OH)HCH_3 \rightarrow ?$$

A. $CH_3CH_2CH_2COOCH(CH_3)_2$

B. $CH_3CH_2CH_2CH_2C(OH)OCH(CH_3)_2$

C. $CH_3CH_2CH_2CH_2COOCHCH_2CH_3$

D. $CH_3CH_2CH_2CH_2COOCH(CH_3)_2$

27. The products of acid hydrolysis of an ester are:

A. alcohol + water

B. acid + water

C. another ester + water

D. alcohol + acid

28. Which organic functional group is important for its basic properties?

A. carbonyl

B. hydroxyl

C. amine

D. aromatic

29. Give the formula of the structure below:

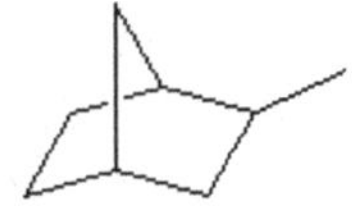

A. C_8H_{14}

B. C_8H_{12}

C. C_8H_{10}

D. C_8H_8

---

30. Which of the following structures, including formal charges, is correct for diazomethane, CH_2N_2?

A. $H_2C=N^+=N^-$

B. $H_2C=N^+\equiv N^-$

C. $H_2C=N^-=N^+$

D. $^-CH_2-N\equiv N:$

31. How many stereoisomers are possible for the structure below?

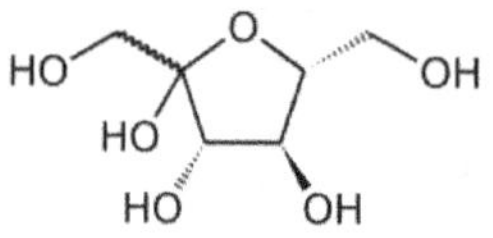

A. 2   **B.** 4   **C.** 8   **D.** 16

32. Which compound is expected to show intense IR absorption at 1715 cm^{-1}?

A. $(CH_3)_2CHNH_2$

B. hex-1-yne

C. 2-methylhexane

D. $(CH_3)_2CHCO_2H$

33. If the concentration of ^-OH doubles in a reaction with bromopropane, then the reaction rate:

A. quadruples

B. doubles

C. remains the same

D. is halved

34. Give the best product for the following reaction:

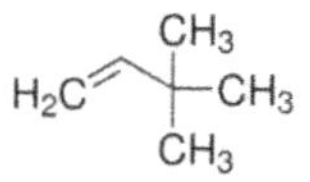

$$H_2C\!=\!C(CH_3)\!-\!C(CH_3)_2\!-\!CH_3 \; + \; 1)\; Hg(O_2CCF_3)_2,\; CH_3CH_2OH;\; 2)\; NaBH_4 \rightarrow ?$$

A. $CH_3-CH(CH_3)-C(CH_3)(OCH_2CH_3)-CH_3$

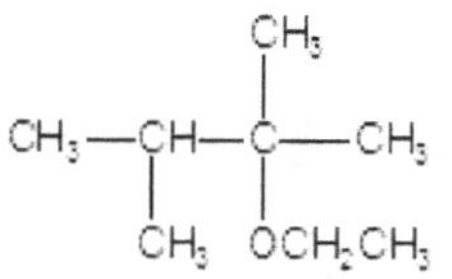

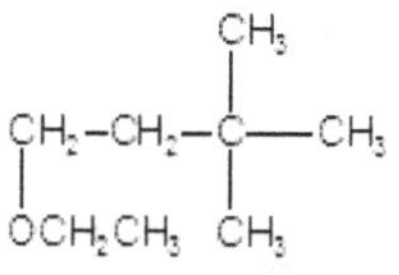

B. $CH_3-CH_2-C(CH_3)(CH_3)-CH_2OCH_2CH_3$

C. $CH_2-CH_2-C(OCH_2CH_3)(CH_3)-CH_3$

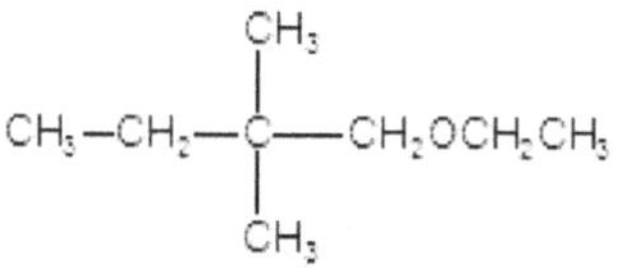

D. $CH_3-CH(CH_3)-C(CH_2CH_3O)(CH_3)-CH_3$

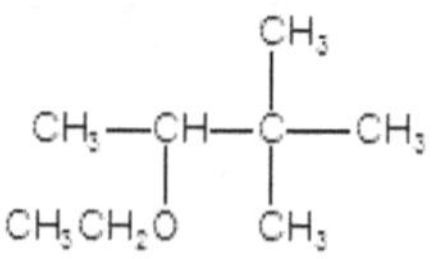

35. Which of the following molecular formulas correspond(s) to an acyclic alkyne?

     I. C_9H_{20} II. C_9H_{18} III. C_9H_{16}

A. I only

B. II only

C. III only

D. I and II only

36. How many pairs of degenerate π molecular orbitals are in benzene?

   A. 6   **B.** 2   **C.** 4   **D.** 3

37. When an alcohol reacts with phosphoric acid, the product is a:

   A. pyrophosphate   **C.** phosphate ester

   B. phosphate anion   **D.** phosphate salt

38. What is an ester reduced to with diisobutylaluminum hydride (DIBAL)?

   A. 1° alcohol   **C.** ketone

   B. alkane   **D.** aldehyde

Notes for active learning

Answer Keys

and

Detailed Explanations

Diagnostic Test 1 – Answer Key and Detailed Explanations

#	Ans	Topic	#	Ans	Topic
1	D	Nomenclature	36	D	Aromatic compounds
2	B	Covalent bond	37	B	Alcohols
3	C	Stereochemistry	38	B	Aldehydes & ketones
4	B	Molecular structure & spectra	39	A	Carboxylic acids
5	C	Separations & purifications	40	B	COOH derivatives
6	B	Alkanes & alkyl halides	41	A	Amines
7	A	Alkenes	42	B	Nomenclature
8	C	Alkynes	43	C	Covalent bond
9	C	Aromatic compounds	44	A	Stereochemistry
10	B	Alcohols	45	A	Nomenclature
11	D	Aldehydes & ketones	46	D	Covalent bond
12	A	Carboxylic acids	47	D	Stereochemistry
13	A	COOH derivatives	48	C	Molecular structure & spectra
14	B	Amines	49	C	Separations & purifications
15	B	Nomenclature	50	D	Alkanes & alkyl halides
16	B	Covalent bond	51	A	Alkenes
17	B	Stereochemistry	52	B	Alkynes
18	B	Molecular structure & spectra	53	D	Aromatic compounds
19	C	Separations & purifications	54	C	Alcohols
20	D	Alkanes & alkyl halides	55	A	Aldehydes & ketones
21	C	Alkenes	56	C	Carboxylic acids
22	C	Alkynes	57	C	COOH derivatives
23	A	Aromatic compounds	58	A	Amines
24	B	Alcohols	59	B	Nomenclature
25	A	Aldehydes & ketones	60	C	Covalent bond
26	D	Carboxylic acids	61	C	Stereochemistry
27	B	COOH derivatives	62	B	Molecular structure & spectra
28	D	Amines	63	C	Separations & purifications
29	C	Nomenclature	64	C	Alkanes & alkyl halides
30	A	Covalent bond	65	A	Alkenes
31	D	Stereochemistry	61	A	Alkynes
32	B	Molecular structure & spectra	67	C	Aromatic compounds
33	C	Alkanes & alkyl halides	68	B	Alcohols
34	A	Alkenes	69	B	Aldehydes & ketones
35	B	Alkynes	70	B	Carboxylic acids

This page is intentionally left blank

1. D is correct.

Number this five-carbon chain beginning at the highest-priority functional group – the alcohol (functional groups with higher oxidation states are a higher priority). Thus, the alcohol is attached to carbon 1.

The C=C double bond is between carbons 2 and 3, and the C≡C triple bond is between carbons 4 and 5.

The stereochemistry about the alkene is '*E*' (highest priority groups are on opposite sides of the alkene).

The priority groups are ranked according to the Cahn-Ingold-Prelog rules for prioritization based on the atomic number of atoms attached to the alkene.

Cis–trans relationship cannot be used to describe the molecule because the substituents across the double bond are different.

2. B is correct.

There are four bonding patterns of carbon described by the three hybridization bonding models.

When carbon forms four single (σ) bonds, it is sp^3 hybridized.

When carbon forms one double (π) bond and two single (σ) bonds, it is sp^2 hybridized.

Carbon is sp hybridized when it forms one triple ($2-\pi$) bond and one single (σ) bond or two double (π) bonds.

$$O=CH-CH_2-CH=C=C=CH_2$$

$$sp^2 \quad sp^3 \quad sp^2 \quad sp \quad sp \quad sp^2$$

3. C is correct.

Inspection of each molecule reveals that the longest chain in both molecules is 10 carbon atoms.

These molecules have the same root name and the substituents, and the locations of the substituents are the same. Therefore, the molecules represent the same compound.

4. B is correct.

Infrared radiation (IR) is useful in the region of 1500 to 3500 cm^{-1} (called wavenumbers).

In this range, the molecular vibrations of molecules are active, and their characteristic absorbance frequencies identify the functional groups.

Below 1500 cm^{-1} is the fingerprint region used for more detailed analysis once the target molecule(s) have been identified.

A: Nuclear magnetic resonance (NMR) spectroscopy involves a sample containing the compound being subjected to a high-intensity magnetic field and scanning through the radio-frequency range of the electromagnetic spectrum for particular absorptions.

NMR relies on the magnetic properties of certain atomic nuclei and determines the physical and chemical properties of atoms within molecules. NMR can be used to deduce the structure (connectivity) of the atoms within the molecule.

C: The UV range (not the IR range) of wavelengths is between 200–400 nm and corresponds to the energy required for electronic transitions between the bonding or nonbonding molecular orbitals and antibonding molecular orbitals.

UV spectroscopy is useful for studying compounds that contain double bonds, especially in conjugated (i.e., alternating double and single bond) molecules.

When molecules containing π-electrons (or non-bonding electrons) absorb UV energy, these electrons are promoted (i.e., excited) to higher *anti*-bonding molecular orbitals.

The more easily excited the electrons (i.e., lower energy gap between the *HOMO* and the *LUMO*), the longer the wavelength of UV light it absorbs.

D: MS (mass spectrometry) studies compounds through the fragmentation of molecules, although the unfragmented parent peak provides useful information.

Note: mass spectrometry destroys (fragments) the sample and is not preferred for rare/limited samples.

5. C is correct.

For thin-layer chromatography (TLC), a mixture is spotted onto the thin absorbent layer, affixed to plastic or glass (polar) plates.

The spotted TLC plate is placed upright in a developing chamber that contains a solvent.

The solvent is typically a nonpolar liquid, such as ether.

Capillary action draws the solvent up the TLC plate and portions of the mixture (sample spotted onto the plate) that are miscible (dissolved) by the solvent.

Solubility (i.e., migrations along with the solvent) of the sample depends on the affinity of the sample to adhere to the polar plates compared to its solubility in the nonpolar (e.g., ether) solvent.

The R_f value (i.e., retention factor) is a physical reference value for that molecule.

The distance traveled by the sample is divided into the distance traveled by the solvent.

The fraction is multiplied by 100% and expressed as a percentage between 0% (highly polar molecules that do not migrate) to 100% (highly non-polar molecules that dissolve and then migrate along with the solvent).

6. B is correct.

Bimolecular nucleophilic substitution (S_N2) occurs at the fastest rate when the substrate is the least hindered.

The relative reactivity of the alkyl halides for S_N2 is methyl > primary > secondary >> tertiary.

S_N2 reactions do not occur on sterically hindered tertiary substrates.

A: 1-chloro-2,2-diethylcyclopentane is a secondary alkyl halide. This substrate does not undergo S_N2 substitution as rapidly as 1-chlorocyclopentane because the presence of branching adjacent to the carbon with the leaving group reduces the rate of S_N2 reactions since the approach of the nucleophile is impeded, compared to a straight-chain molecule.

C: 1-chlorocyclopentene does not undergo nucleophilic substitution because it has a halogen attached to a vinylic (i.e., on a double bond) carbon.

D: 1-chloro-1-ethylcyclopentane is a tertiary halide and does not undergo S_N2 nucleophilic substitution. The mechanism is S_N1 (with a carbocation intermediate).

7. A is correct.

Tertiary alkyl halides form alkenes with strong bases via E_2, such as sodium ethoxide ($NaOCH_2CH_3$).

The most substituted alkene is the major (Zaitsev) product that is internal or more substituted.

The least substituted alkene is the minor (Hofmann) product that is terminal or less substituted.

B: 2-methylpent-3-ene is a molecule whereby the alkene does not connect to carbon that had the bromine or with an adjacent carbon. Rearrangement does not occur in E_2 reactions because rearrangement requires a carbocation (S_N1 or E_1) intermediate.

C: 2-methyl-2-methoxypentane is the product of substitution. Tertiary alkyl halides do not undergo substitution with Lewis bases; elimination is the mechanism for product formation.

D: 2-methylpentene is a less substituted alkene and only occurs with a bulky base (e.g., tert-butyl oxide or LDA).

8. C is correct.

Alkynes can be oxidized to aldehydes or ketones as follows:

$$CH_3C{\equiv}CH \quad \xrightarrow[\text{HgSO}_4]{\text{H}_2\text{O, H}_2\text{SO}_4} \quad CH_3\overset{\displaystyle OH}{\underset{\displaystyle |}{C}}{=}CH_2 \;\rightleftharpoons\; CH_3\overset{\displaystyle O}{\overset{\|}{C}}CH_3 \quad \text{a ketone}$$

$$CH_3C{\equiv}CH \quad \xrightarrow[\text{2. HO}^-,\ \text{H}_2\text{O}_2,\ \text{H}_2\text{O}]{\text{1. disiamylborane}} \quad CH_3CH{=}\overset{\displaystyle OH}{\underset{\displaystyle |}{C}}H \;\rightleftharpoons\; CH_3CH_2\overset{\displaystyle O}{\overset{\|}{C}}H \quad \text{an aldehyde}$$

Disiamylborane (BH_3) is used for the hydroboration of alkynes (and alkenes) and involves peroxides in step 2.

The hydration of the alkyne (or alkene) proceeds through an *anti*-Markovnikov addition; the preference for the *anti*-Markovnikov addition is due to the minimization of steric interactions.

The peroxide (H_2O_2) and ^-OH converts the RBH_2 bond to an enol (C=C–OH) of the *anti*-Markovnikov product.

The enol (alkene and alcohol attached to the same carbon atom) is less stable and tautomerizes (i.e., migration of a proton) to the aldehyde (aldehyde or ketone are referred to as keto) and is not the final product of the reaction.

H

O

tautomerism

O

H

H

enol form   keto form

The keto and enol form are structural isomers, with the keto form more than 99% of the final yield due to stability.

9. C is correct.

Each molecule has nitro (~NO_2) and hydroxyl (~OH) functional groups that can hydrogen-bond.

Proximity is needed for either intramolecular (within the molecule) or intermolecular (between molecules) bonding. The alignment of the functional groups to permit hydrogen bonding depends on the shape of the individual molecule.

The melting points (transition from packed molecules in a solid to liquid phase), boiling points (transition from associated molecules in a liquid to independent molecules in the gas phase), and water solubility of polar molecules are related to the presence and quantity of intermolecular hydrogen bonding.

The *meta-* and *para*-nitrophenol are more water-soluble and have higher melting points than *ortho*-nitrophenol because *ortho*-nitrophenol tends to form intramolecular hydrogen bonds instead of intermolecular hydrogen bonds.

Intramolecular hydrogen bonds for *ortho*-nitrophenol make the molecule independent so that it takes less energy to disrupt the lattice structure of the molecules (melting). Likewise, intramolecular hydrogen bonding reduces the molecule's ability to form hydrogen bonds with water, and therefore the molecule is less water-soluble.

A: *meta-* and *para*-nitrophenol form strong intermolecular hydrogen bonds, which leads to a higher melting point. The nitro and hydroxyl groups form hydrogen bonds with water, and the molecules are more water-soluble.

B: *ortho*-nitrophenol forms some intermolecular hydrogen bonds, but less than *meta-* and *para*-nitrophenol. *Meta*-nitrophenol forms weak intramolecular bonds due to the large distance between the functional groups.

D: *para*-nitrophenol has the nitro and hydroxyl substituents at opposite ends of the flat molecule and cannot form intramolecular hydrogen bonds.

10. B is correct.

The reaction of (R)-2-hexanol and PBr$_3$ (phosphorous tribromide) proceeds via an S$_N$2 reaction mechanism.

An addition-elimination sequence occurs in the P–Br bond, then substitution by Br$^-$ of the alcohol gives the inverted (S)-2-bromohexane product.

11. D is correct.

The correct tautomer forms of the ketone have the carbon-oxygen bond directly bound to the alkene carbon atom.

12. A is correct.

In Brønsted-Lowry theory, an acid is a proton donor, and a base is a proton acceptor.

In solution, a strong acid dissociates its proton and exists predominantly in the deprotonated form as the acid's conjugate base. A strong acid forms a weak conjugate base because the acid's protons dissociate in solution.

The resulting anions of the deprotonated carboxylic acids are stable due to resonance. Resonance is a major contributor to stabilize the anion by delocalizing the negative charge (using *pi* bonds) over several (in this example, two oxygen) atoms.

Acids that contain an electron-withdrawing substituent on the α-carbon (i.e., carbon adjacent to the carbonyl) tend to be stronger (donate proton more readily) because induction (via *sigma* bond) pulls electron density.

Induction has a stabilizing effect on the carboxylate anion (the conjugate base of carboxylic acid). Conversely, acids with electron-donating α substituents (i.e., methyl chains) tend to be weaker (less likely to dissociate H$^+$).

The acid with two electron-withdrawing chlorine substituents forms the most stable carboxylate anion (when H$^+$ dissociates). Therefore, it is the strongest acid, meaning it has the weakest (most stable) conjugate base.

B: $CH_3CH_2CH_2CO_2H$ is a carboxylic acid with only hydrogens and therefore lacks stability influences from electronegative atoms (F, N, O or Cl).

C: $(CH_3CH_2)_3CCO_2H$ is a carboxylic acid with a tertiary butyl substituent electron-donating via hyperconjugation (i.e., alkyl chains donate electrons).

D: $CH_3HNCH_2CH_2CH_2CO_2H$ is a carboxylic acid with an amino substituent strongly electron-withdrawing via electronegativity of the nitrogen, but the nitrogen is at a large distance from the COO$^-$ and therefore, its influence is minimal.

13. A is correct.

The reaction is the acid hydrolysis of an ester. When esters are hydrolyzed, they yield carboxylic acids (~COOH) and alcohols (~OH).

From the hydrolysis of this ester, 2-butenoic acid and isobutanol are formed.

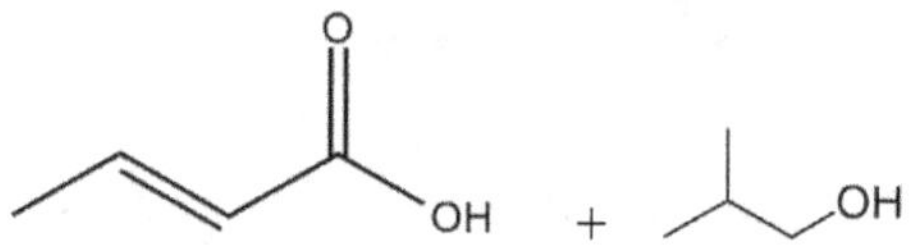

2-butenoic acid   *isobutanol*

B: $CH_3CH(CH_3)_2$ is an alkane, and neither type of compound can be obtained by hydrolyzing an ester.

C: $HOOCCH_2CH(CH_3)_2$ may look like the carboxylic acid produced from the hydrolysis of the ester, but the carbonyl double bond is absent.

D: $CH_3CH=CHCHO$ is an unsaturated aldehyde.

14. B is correct.

Exposing bulky amines to an acid such as HCl allows it to become more soluble in water.

The charged ammonium cation allows for stronger dipole interactions with water compared to the neutral amine form.

15. B is correct.

The longest carbon chain in the molecule contains six carbon atoms.

The highest (and only) functional group is the amide.

16. B is correct.

Full arrowheads show the movement of a pair of electrons (compared to single-headed – fishhook – arrows for radical reactions).

Only movement of the *pi* (π) electrons is responsible for the stable diene structures to the right.

The *pi* electrons must be in conjugation.

17. B is correct.

When determining whether an alkene is the *Z* / *E* (or *cis* / *trans*) stereoisomer, it is important to identify the higher priority substituent group at each of the two carbon atoms of the alkene.

The priority groups are ranked according to the Cahn-Ingold-Prelog (CIP) rules for prioritization based on the atomic number of atoms attached to the alkene.

If the higher priority groups are positioned on the same side of the double bond, the molecule is *Z* (*cis* notation can be used if the substituents are the same).

If the higher priority groups are positioned on the opposite sides of the double bond, the molecule is *E* (*trans* notation can be used if the substituents are the same).

The stereochemistry about the alkene is '*Z*.' The highest priority groups – Br and Cl across the double bond – are on the same side of the alkene.

Cis–trans relationship cannot describe the molecule because the substituents across the double bond are different.

18. B is correct.

The IR absorbance at 1710 cm^{-1} indicates the carbonyl group of either a ketone or an aldehyde. The carbonyl is present in carboxylic acid derivatives: acyl halide, anhydride, carboxylic acid, esters, and amide.

NMR can distinguish between an aldehyde (NMR δ 9–10) and a ketone (no characteristic NMR signals).

Aldehydes have additional peaks at 2700–2800 cm^{-1}, while ketones do not.

19. C is correct.

Because of the significant difference in molecular weight between acetone and octane, they have different boiling points.

Acetone is a 3-carbon chain with a BP of 56–57 °C, while octane is an 8-carbon chain with a BP of 125–126 °C. The approximate magnitude of the difference (not the exact values) in BP is necessary to solve this problem.

In general, a significant increase in molecular weight between two compounds results in a significant difference in boiling point, and therefore distillation separates the molecules.

Distillation is effective for separating molecules with differences in BP of at least 25 °C.

20. D is correct.

Stable molecules are the best leaving groups.

If the leaving groups are charged, then the bromide ion is the best leaving group because the anion is stable (due to atomic size).

The order of the halogens as leaving groups: I$^-$ > Br$^-$ > Cl$^-$ > F$^-$

The order of the halogens as nucleophiles: I$^-$ > Br$^-$ > Cl$^-$ > F$^-$

21. C is correct.

An *anti*-Markovnikov addition of water is needed across the alkene double bond.

The reagents are borane, followed by hydrogen peroxide and sodium hydroxide.

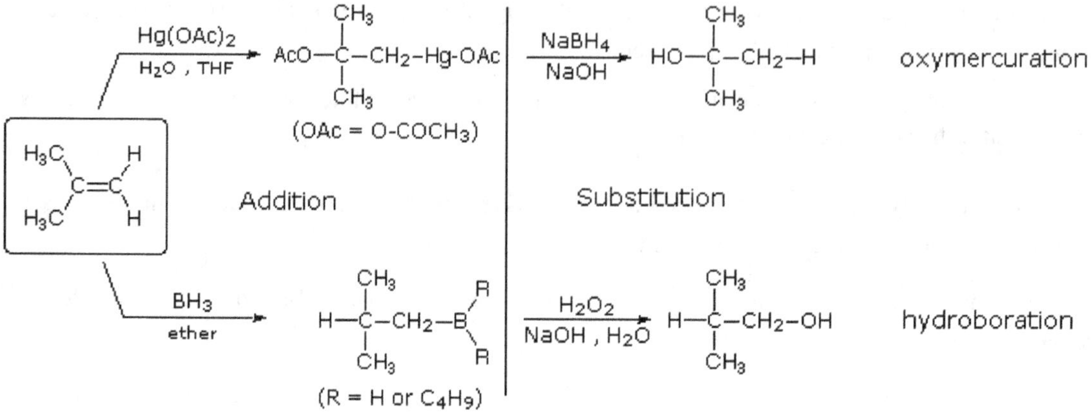

Comparison of Markovnikov vs. anti-Markovnikov addition reactions for alkenes

Each reaction above occurs in separate steps, as indicated by the vertical separation line.

A: the product of oxymercuration-demercuration: 1) $Hg(OAc)_2$, H_2O / THF; 2) $NaBH_4$

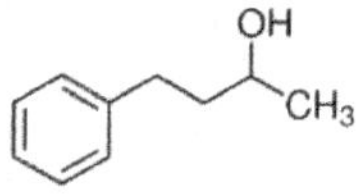

22. C is correct.

Like alkenes, alkynes are electron-rich functional groups and act as nucleophiles that donate electron density to electron-seeking electrophiles.

The reactivity of alkenes and alkynes is similar, and they interact with electrophiles in analogous ways.

23. A is correct.

Hückel's rule predicts that for a monocyclic compound to be aromatic, there must be a fully conjugated *pi* (sp^2 hybridization at each atom in the ring) containing ($4n + 2$) *pi* electrons.

Each of the double bonds contributes two *pi* electrons. The lone pair of electrons on a double-bonded N is perpendicular to the *pi* cloud and does not count as the number of *pi* electrons for Hückel's rule. Note: when N is in a single bond, the unshared electrons are parallel to the *pi* system and count toward aromaticity.

Benzimidazoline is not fully conjugated because there is a CH_2 (sp^3) group in the ring. Cyclic conjugation is necessary for a molecule to be aromatic.

B: thiophene has a sp^2 hybridized sulfur (like oxygen) and has a lone pair of electrons counted in the ring. The number of *pi* electrons is 6.

C: quinoline has 10 *pi* electrons and therefore is aromatic. The lone pair of electrons on a double-bonded N is perpendicular to the *pi* cloud and does not count as the number of *pi* electrons for Hückel's rule.

D: thiazole has *6 pi* electrons and therefore is aromatic. The lone pair of electrons on a double-bonded N is perpendicular to the *pi* cloud and does not count as *pi* electrons for Hückel's rule. Sulfur (like oxygen) is sp^2 hybridized and has a lone pair of electrons counted in the ring.

24. B is correct.

Esterification occurs when a carboxylic acid reacts with alcohol under catalytic acidic conditions to form an ester + water.

A: C_6H_5OH and CH_3CH_2Br form an ether via Williamson (S_N2) ether synthesis when the alcohol reacts with the alkyl halide.

C: $CH_3COOH + SOCl_2$ is the S_N2 reaction of a carboxylic acid with thionyl chloride ($SOCl_2$), and an acyl halide is formed.

D: 2 $CH_3OH + H_2SO_4$ forms dimethyl ether from methanol and catalytic acid (e.g., sulfuric acid).

25. A is correct.

Carboxylic acid derivatives are similar in structure to ketones and aldehydes.

However, one of the H or R groups has been replaced with a heteroatom, such as oxygen or nitrogen.

26. D is correct.

Hydrogen bonding is a strong type of dipole-dipole interaction and raises the boiling point of organic compounds.

Dimer formed from hydrogen bonding of two carboxylic acids

The carboxylic acid of acetic acid participates in hydrogen bonding, while the ester of methyl acetate does not form similar intermolecular hydrogen bonds.

27. B is correct.

Acid bromides have the best leaving group (i.e., most stable anion), and therefore are the most reactive. Acyl halides (RCOCl or RCOBr) are so reactive that they are not typically found in nature due to the moisture in the atmosphere, which converts them to carboxylic acids.

The reaction rate depends on the leaving group, the steric environment, and the substituents bonded to the carbonyl carbon.

The more stable the leaving group (as an anion), the faster the reaction is. A more electrophilic carbonyl (i.e., induction from neighboring groups) undergoes nucleophile attack more rapidly.

Additionally, steric hindrance (i.e., bulky groups) at the reaction center leads to slower reaction rates.

28. D is correct.

A tertiary amine (R_3N) has three alkyl groups bonded to the nitrogen atom.

A secondary amine (R_2NH) has two alkyl groups bonded to the nitrogen atom.

A primary amine (RNH_2) has one alkyl group bonded to the nitrogen atom.

A and C: are primary amines.

B: is a secondary amine.

29. C is correct.

Neutral carbon atoms maintain bonds to four other atoms; therefore, an acyclic hydrocarbon cannot terminate with a methylene ($\sim CH_2\sim$) group.

Methyl ($\sim CH_3$) groups are at the ends of alkanes and have a formula of $\sim CH_3$.

Using a subscript of n for the number of carbons, the degrees of unsaturation can be determined from the following formulae:

       Alkane: C_nH_{2n+2} = 0 degrees of unsaturation

       Alkene: C_nH_{2n} = 1 degree of unsaturation

       Alkyne: C_nH_{2n-2} = 2 degrees of unsaturation

$CH_3CH_3CH_3$ has 3 carbons and, according to the formula C_nH_{2n+2}, should have 8 hydrogens. This molecule has 9 hydrogens; it is impossible because it would require a carbon with 5 bonds.

A: $CH_3CHCH_3CH_2CH_3$ has 5 carbons and, according to the formula C_nH_{2n+2}, should have 12 hydrogens.

B: $CH_3CH_2CH_2CH_2CH_3$ has 5 carbons and, according to the formula C_nH_{2n+2}, should have 12 hydrogens.

D: $CH_3CH_2CH_2CH_3$ has 4 carbons and, according to the formula C_nH_{2n+2}, should have 10 hydrogens.

30. A is correct.

There are four regions of electron density around the nitrogen atom (including the lone pair). Therefore, the nitrogen atom is sp^3 hybridized.

Electron geometry describes the geometry of the electron pairs, groups, and domains on the central atom, whether they are bonding or non-bonding. Molecular geometry is the name of the shape used to describe the molecule. When atoms bond to a central atom, they do it to maximize the distance between bonding electrons, which gives the molecule its overall shape. If no lone pairs of electrons are present, the electronic geometry is the same as the molecular shape. A lone pair occupies more space than bonding electrons, so the net effect is to bend the molecule's shape (although the electron geometry conforms to the predicted shape).

The shape of this molecule is trigonal pyramidal with bonds of approximately 109.5 degrees due to the bulky ethyl substituents. The presence of three substituents (i.e., ethyl) and the lone pair on the nitrogen result in the pyramidal shape consistent with VESPER theory.

31. D is correct.

Determine the molecular formula of the molecule. 2-methylbutane has 5 carbon atoms and 12 hydrogen atoms.

Only *n*-pentane has the same molecular formula.

The notation *n–* represents *normal* (or straight chain).

32. B is correct.

The broad, deep absorption between 3000 cm^{-1} to 3500 cm^{-1} is characteristic for alcohol.

33. C is correct.

Radical termination steps involve an overall decrease in the number of radicals when comparing the starting materials to the products.

The correct answer involves a decrease in the number of radicals (from two to zero).

A: the number of radicals increases from zero to two, which is an initiation step.

B: the number of radicals (one) does not change, which are propagation steps.

C: the number of radicals decreases from two to zero, which is a termination step.

D: the number of radicals (one) does not change, which are propagation steps.

34. A is correct.

The *Cope elimination* is an intramolecular elimination reaction that occurs when the oxygen atom of the oxide of a tertiary amine removes a proton from an adjacent position.

This results in the formation of a *syn*-alkene.

The reaction requires heat.

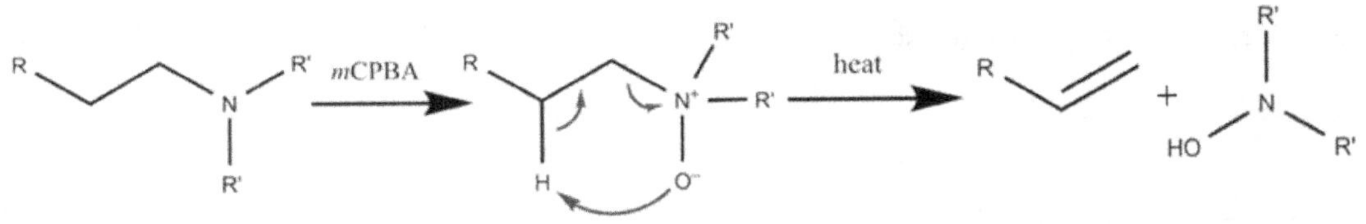

The *Cope elimination* yields the same products as a *Hofmann elimination* (i.e., exhaustive methylation)

35. B is correct.

In this reaction, the bromine adds to the alkene group of C_2H_4 and forms 1,2–dibromoethane.

The reaction proceeds via *anti*-addition from the 3-membered bridge structure of the bromonium (i.e., halonium) ion.

A: hydrogen gas cannot be generated as a product from this reaction.

Carbon-carbon multibonds do not form from the given reaction conditions because the reagents do not include any base to eliminate the bromine(s) to form either an alkene or alkyne.

Sample anti-stereochemical products from the addition of bromine to an alkene

Each product has an enantiomer that is not shown.

Bromonium ion as an intermediate:

bromonium ion

The reaction is regioselective (i.e., where) for the addition of the second nucleophile (i.e., Br^-) to the halonium (i.e., bromonium) structure.

The incoming nucleophile attacks the more substituted atom.

Mechanism of Br_2 addition to an alkene:

bridged halonium ion

The incoming nucleophile attacks the bromonium bridged structure ion at the most substituted position.

The bromonium ion (i.e., 3-membered bridged structure) undergoes S_N2 attack for *anti*-addition product formation.

The stereochemical (i.e., *trans*) notation would include wedges and dashes.

trans-1,2 dibromocyclohexane

36. D is correct.

Degrees of unsaturation:

> double bonds = 1 degree of unsaturation

> triple bonds = 2 degrees of unsaturation

> rings = 1 degree of unsaturation

Using the degree of unsaturation calculation:

> benzene (1 ring and 3 double bonds) = 4 degrees of unsaturation

37. B is correct.

The hydrobromic acid protonates the secondary alcohol that dissociates as water.

The secondary carbocation, formed as an intermediate, is repositioned to the tertiary position through an alkyl (i.e., methide) shift.

After the methide shift, this more stable tertiary carbocation is then attacked by the bromide ion.

38. B is correct.

A ketone is being converted to an alkene.

The oxidation state of ketones is larger than the oxidation state of alkenes, and therefore the ketone needs to be reduced.

Reduction of ketone forms secondary alcohol.

The carbon skeleton rearranges, and this requires the formation of a carbocation.

Exposure to phosphoric acid produces a secondary carbocation, and an alkyl shift and deprotonation follow this to yield the alkene product.

39. A is correct.

The pK_a is the pH where half of the acid has dissociated to its conjugate base form.

The acid and conjugate base concentration is equal when the pK_a of the compound and the pH of the solution are the same.

Henderson-Hasselbalch equation:

> $pH = pK_a + \log[\text{conjugate base}] / [\text{acid}]$

40. B is correct.

The hydrolysis of an ester is the reverse process of condensation because water is introduced to the ester to produce carboxylic acid and alcohol.

A chemical equilibrium exists for the acid-catalyzed process, and the reaction is driven forward using a large excess of water.

A chemical equilibrium does not exist for the base-promoted process because the alcohol product cannot add into the carboxylate to reform the ester.

41. A is correct.

Structure 1: cyclic secondary enamine

Structure 2: secondary amine

Structure 3: cyclic quaternary amine

The secondary amine (2) contains one N–H bond, and therefore, this amine can form hydrogen bonds.

The cyclic amine (3) and the enamine (1) do not contain N–H bonds, so their boiling points are lower.

Aside from hydrogen bonding considerations, molecules with formal charges (molecule 3) interact via electrostatic forces, which increase their boiling point.

42. B is correct.

Because the group has three carbon atoms, the group is a *propyl* group.

Propyl substituents exist as either the *n*-propyl (i.e., *normal* or straight chain) or the isopropyl group.

<table>
<tr><td align="center">isopropyl</td><td align="center">sec-butyl</td><td align="center">tert-butyl</td><td align="center">isobutyl</td></tr>
</table>

Sample common names for organic substituents used in nomenclature

43. C is correct.

Tertiary carbocations are more stable than secondary or primary carbocation because they experience more hyperconjugation effects from the neighboring C–H bonds.

Resonance stabilization lowers the energy of the cation.

44. A is correct.

For an alkene to experience *cis-trans* isomerization, the *pi* bond of the double bond is broken.

The *pi* bond of the alkene makes the double bond rigid.

Heating the alkene to hot temperatures or exposure to electromagnetic radiation (e.g., UV radiation) may cause the *pi* bond to homolytically cleave to 1,2-diradical and rotate about the *sigma* bond to form the diastereomers.

E / *Z* and *cis* / *trans* isomers are geometric isomers and are classified as diastereomers.

45. A is correct.

Draw the line formula of the carbon chain with proper interpretation of the subscripts.

$$H-\underset{\underset{H}{|}}{\overset{\overset{H}{|}}{C}}-\underset{\underset{H}{|}}{\overset{\overset{H}{|}}{C}}-\underset{\underset{H}{|}}{\overset{\overset{H}{|}}{C}}-\underset{\underset{H}{|}}{\overset{\overset{H}{|}}{C}}-\underset{\underset{H}{|}}{\overset{\overset{H}{|}}{C}}-\underset{\underset{H}{|}}{\overset{\overset{H}{|}}{C}}-\underset{\underset{H}{|}}{\overset{\overset{H}{|}}{C}}-H$$

There are two methyl groups and five *methylene* ($\sim CH_2\sim$) in the molecule for a total of seven carbons.

Use the following formula to calculate the *degrees of unsaturation*:

C_nH_{2n+2}: for an alkane (0 degrees of unsaturation)

46. D is correct.

Except for structure D, the other structures are intermediates for electrophilic aromatic substitution (EAS).

These resonance intermediates typically have one cation and no negative charges in the ring.

Resonance hybrids of aniline during electrophilic aromatic substitution

47. D is correct.

There are only two ways to draw butane, either straight-chain or the tertiary alkyl constitutional isomer. The term "*constitutional isomers*" is recognized by IUPAC, while "*structural isomers*" was used historically.

Two constitutional isomers of n-butane (left) and isobutene (right)

48. C is correct.

1-chlorobutane (below) has a chiral center and produces four (4) NMR peaks.

A: 3,3-dichloropentane (below) is a symmetrical molecule that produces two (2) NMR proton peaks.

B: 4,4-dichloroheptane (below) is a symmetrical molecule that produces three (3) NMR proton peaks.

D: 1,4-dichlorobutane (below) produces two (2) NMR proton peaks.

49. C is correct.

The extraction with sodium bicarbonate ionizes the carboxylic acid by deprotonating it to its carboxylate form.

This negatively charged anion has a greater affinity for the aqueous layer.

50. D is correct.

Unimolecular elimination occurs via E_1 and involves forming a carbocation in the slow (rate-determining) step.

A: a single step (concerted) process describes bimolecular (E_2), not E_1 elimination.

B: *homolytic* (compared to the more common *heterolytic*) cleavage of a covalent bond yields free radicals.

C: *free radicals* (i.e., atom with unpaired electron) occur with peroxides (H_2O_2) or dihalides / UV, and radicals are an intermediate in unimolecular (E_1) elimination.

51. A is correct.

A *conjugated* system contains two or more double or triple bonds separated by a single bond (i.e., sp^2 hybridization at each carbon in the conjugated system).

1,2-butadiene is not conjugated because the double bonds are adjacent (not separated by a single bond).

Cumulated double bonds are adhacent and unstable.

B: cyclobutadiene is conjugated because the molecule has two double bonds separated by single bonds.

C: benzene is conjugated because a single bond separates each double bond.

D: 1,3-cyclohexadiene is conjugated because a single bond separates the double bonds between carbons 1-2 and carbons 3-4.

52. B is correct.

The substrate contains a carbon-carbon triple bond at the end of the chain.

Therefore, to synthesize 2-hexanone from 1-hexyne, oxygen is introduced to the internal carbon atom of the alkene. This reaction requires aqueous acidic conditions so that the addition proceeds as a Markovnikov addition.

The protonation of the alkyne results in a secondary vinyl cation. This cation is trapped by water to produce an enol that tautomerizes to form the ketone product.

53. D is correct.

The methyl group is an *electron-donating* substituent and therefore is an *ortho / para* director.

Due to steric hindrance at the *ortho* position, the formation of the *para* product is the major product.

54. C is correct.

Carboxylic acids are functional groups with an ~OH group bonded to carbonyls (C=O).

Oximes C=N–OH are functional groups containing an ~OH group; used for reductive amination.

55. A is correct.

The treatment of a terminal alkyne with a dialkyl borane results in a vinyl borane.

The vinyl borane is oxidized to the enol by hydrogen peroxide (H_2O_2) and the hydroxide ($^-$OH) base.

This group tautomerizes to form an aldehyde.

56. C is correct.

The reaction used to combine alcohols and carboxylic acids is the Fischer esterification reaction.

$$CH_3\text{-}C(\!=\!O)O\text{-}H + CH_3CH_2OH \;\leftrightarrow\; CH_3\text{-}C(\!=\!O)O\text{-}CH_2CH_3 + H_2O$$

Acid catalysis is needed to activate the carboxylic acid carbonyl group for the nucleophilic attack.

57. C is correct.

An amide is an amine directly bonded to a carbonyl (~C=O) group.

The carbonyl group accepts the lone pair of electrons on the nitrogen atom through resonance action.

Amides ($pK_a \approx 20\text{–}25$) are less electrophilic than esters.

Primary and secondary amides have acidic N–H protons ($pK_a \approx 35$).

58. A is correct.

Boiling points of compounds are determined by two factors: *molecular weight* and *intermolecular interactions*.

The higher the molecular weight, the harder it is to "push" it into the gas phase, and the higher the boiling point.

Similarly, the stronger the intermolecular interactions, the more energy is required to disrupt them and separate the molecules in the gas phase, hence the higher the boiling point.

Dimethylamine has one N–H bond (donor) and one lone pair on the nitrogen (acceptor):

$$H_3C\text{—}N(H)\text{—}CH_3$$

The other answer choices can accept and donate hydrogen bonds, which increases their boiling points.

59. B is correct.

The longest carbon chain is composed of five carbon atoms.

There are two alkenes, so the root name is "pentadiene."

There is a methyl group in the second position along the carbon chain.

60. C is correct.

The bonds between atoms (i.e., intramolecular) are stronger than the bonds between molecules (i.e., intermolecular).

Examples of intermolecular bonds are hydrogen, dipole-dipole, dipole-induced dipole, and van der Waals.

Sigma bonds are single bonds and involve overlap along the internuclear axis between the atoms.

S*igma* bonds are stronger than the *pi* bond of a double bond.

A: *hydrogen bonding* is a common intramolecular bond in which a hydrogen atom of one molecule is attracted to an electronegative atom (nitrogen, oxygen, or fluorine).

B: dipole-dipole bonds are attractive forces between the positive end of one polar molecule and the negative end of another polar molecule.

D: *ionic bonds* result from the complete valence electron(s) transfer between atoms; it generates two oppositely charged ions. The metal loses electrons to become a positively charged cation, whereas the nonmetal accepts those electrons to become a negatively charged anion.

Ionic bonds can be disrupted in water and weaker in aqueous solutions than in a dry environment.

61. C is correct.

Geometric (configurational) isomers have the same molecular formula but different connectivity between the atoms due to the orientation of substituents around a carbon-carbon double bond (or ring).

In Z (*cis* when substituents are the same) isomers, the same substituents are on one side of the double bond or ring, while in E (*trans* when substituents are the same) isomers, the same substituents are on opposite sides of the double bond or ring.

62. B is correct.

Carboxylic acid protons are some of the most deshielded protons.

These functional groups can partially ionize, causing the hydrogen to develop a high positive (cationic) character.

63. C is correct.

Molecules with greater molecular mass and that experience stronger intermolecular forces (e.g., hydrogen bonding) tend to have higher boiling points.

These compounds are less volatile and are near the bottom of the fractioning tower.

64. C is correct.

Since an S_N1 reaction proceeds through a planar, sp^2 hybridized carbocation intermediate (which can be attacked from either side), it forms a racemic mixture (i.e., both R / S stereoisomers are present).

65. A is correct.

In $(CH_3)_2C=C(CH_3)CH_2CH_3$, both carbons of the alkene are equally substituted (i.e., tertiary), so the two putative carbocations are approximately equally stable.

Adding a hydrogen halide across a double bond can create two chiral centers, at each of the former sp^2 carbons.

The carbocation forms preferentially at the more substituted carbon.

66. A is correct.

Radical hydrogenation of a C≡C triple bond in the presence of sodium metal and liquid ammonia adds two hydrogen atoms (i.e., reduction) across the double bond with *anti*-stereoselectivity, producing an E alkene.

The mechanism for reducing an alkyne to a trans (E) alkene

Lindlar reagents (i.e., H_2, Pd, $CaCo_3$, quinolone, and hexane) reduce the alkyne to the *cis* (Z) alkene.

The reaction for reducing an alkyne to a cis (Z) alkene

The Lindlar reagent, unlike the reagents of Na and NH_3, can be used to reduce a terminal alkyne to an alkene. Na and NH_3 is a "poison catalyst" and are unreactive with a terminal alkyne.

67. C is correct.

The unknown cyclic hydrocarbon does not react with bromine in dichloromethane, carbon tetrachloride, or water. This means it cannot contain non-conjugated double or triple bonds; otherwise, the bromine would have added across the double bonds.

Since the unknown compound reacts when $FeBr_3$ is added, it must be benzene. The $FeBr_3$ acts as a Lewis acid and catalyzes aromatic electrophilic addition reactions.

Bromine is not a strong enough electrophile to disrupt the conjugated double bonding in the benzene ring. However, adding the iron (III) bromide catalyst converts the bromine into a strong enough electrophile that adds to the benzene ring.

68. B is correct.

The reaction between (S)-2-heptanol and $SOCl_2$ (thionyl chloride) proceeds via an S_N2 reaction mechanism.

An addition-elimination sequence occurs at the S=O double bond, then substitution by Cl⁻ of the alcohol gives the inverted (R)-2-chloroheptane product.

69. B is correct.

The *iodoform reaction* involves trihalogenating the methyl group to turn it into a good leaving group.

After tribromination of the methyl group has occurred, the group is expelled from the molecule by adding hydroxide to the ketone.

Molecule 4 undergoes tribromination before being replaced by the ⁻OH

Acidic conditions only provide the monobrominated compound from the starting ketone; therefore, basic conditions must be used during the reaction.

70. B is correct.

Carboxylic acids have some of the highest boiling points due to hydrogens for dimeric structures.

Alcohols contain the polar O–H bond and form hydrogen bonds, thus contributing to their high boiling point.

Notes for active learning

Notes for active learning

Diagnostic Test 2 – Answer Key and Detailed Explanations

1	A	Nomenclature	36	B	Aromatic compounds
2	B	Covalent bond	37	C	Alcohols
3	D	Stereochemistry	38	B	Aldehydes & ketones
4	B	Molecular structure & spectra	39	B	Carboxylic acids
5	C	Separations & purifications	40	B	COOH derivatives
6	D	Alkanes & alkyl halides	41	D	Amines
7	B	Alkenes	42	C	Nomenclature
8	D	Alkynes	43	A	Covalent bond
9	A	Aromatic compounds	44	B	Stereochemistry
10	D	Alcohols	45	D	COOH derivatives
11	D	Aldehydes & ketones	46	C	Amines
12	C	Carboxylic acids	47	C	Nomenclature
13	B	COOH derivatives	48	D	Covalent bond
14	C	Amines	49	A	Stereochemistry
15	A	Nomenclature	50	C	Molecular structure & spectra
16	D	Covalent bond	51	B	Alkanes & alkyl halides
17	A	Stereochemistry	52	C	Alkenes
18	A	Molecular structure & spectra	53	C	Alkynes
19	C	Separations & purifications	54	D	Aromatic compounds
20	B	Alkanes & alkyl halides	55	A	Alcohols
21	A	Alkenes	56	A	Aldehydes & ketones
22	A	Alkynes	57	D	Carboxylic acids
23	A	Aromatic compounds	58	B	COOH derivatives
24	D	Alcohols	59	D	Amines
25	C	Aldehydes & ketones	60	B	Nomenclature
26	D	Carboxylic acids	61	A	Covalent bond
27	B	COOH derivatives	62	C	Stereochemistry
28	A	Amines	63	D	Nomenclature
29	C	Nomenclature	64	D	Covalent bond
30	C	Covalent bond	65	A	Stereochemistry
31	B	Stereochemistry	66	C	Molecular structure & spectra
32	D	Molecular structure & spectra	67	B	Separations & purifications
33	A	Alkanes & alkyl halides	68	B	Alkanes & alkyl halides
34	A	Alkenes	69	C	Alkenes
35	C	Alkynes	70	C	Alkynes

This page is intentionally left blank

1. A is correct.

The suffix ~*oate* signifies an ester functional group as the highest priority group in the molecule.

2. B is correct.

The carbonyl carbon is trigonal planar because the double-bonded carbon is sp^2 hybridized.

I: each of the two methyl carbons is sp^3 hybridized and tetrahedral.

III: none of the carbons have an unshared pair of electrons (i.e., carbanions) because carbanions are highly reactive and observed in a limited number of examples (e.g., Grignard reagent, Gillman reagent, acetylide anion) and are not present in stable molecules.

3. D is correct.

A *meso* compound has chiral centers but is not chiral because it has an internal plane of symmetry.

A *meso* compound is a single molecule that contains two enantiomers joined together.

Tartaric acid is a four-carbon polyol with two carboxylic acids and two chiral centers. There are three stereoisomers: the (+) form, the (–) form, and the *meso* form.

Meso compounds are identical to their mirror images.

Each of the two stereoisomers rotates plane-polarized light as indicated by the notation of (+) or (–), but *meso*-tartaric acid is achiral (i.e., no net rotation of plane-polarized light).

III: racemic mixture refers to a solution that contains enantiomers (i.e., chiral molecules that are mirror images).

4. B is correct.

For NMR, the position that hydrogen absorbs is determined by the chemical environment.

If the chemical environment of two hydrogens is identical, only one signal is produced. Therefore, equivalent hydrogens could be replaced by another group to yield the same molecule.

The challenge is in determining whether hydrogens are in identical environments (i.e., symmetric molecules).

1,2-dibromoethane

1,2-dibromoethane produces only one absorbance in the NMR spectrum because a molecule has only one type of hydrogen atom.

A: *tert*-butyl alcohol produces 2 signals.

C: toluene produces 2 signals.

D: methanol produces 2 signals.

5. C is correct.

A molecule must have a net positive charge to migrate to the cathode.

In gel electrophoresis, the cathode is the negatively charged end because it involves an electrolytic rather than a galvanic cell.

Amino acids have amphoteric properties ($+$ / $-$ regions within the molecule) at physiological pH of 7.35; the carboxylic acid end is negative, while the amino end is positive.

At physiological pH values, the carboxyl terminus of amino acid with COOH ($pK_a \approx 5$) dissociates into a negatively charged carboxylate.

At physiological pH, the amino terminus with NH_3 ($pK_a \approx 10$) is protonated and becomes positively charged.

Thus, at physiological pH, an amino acid is a zwitterion (i.e., both positive and negative charges within the same molecule).

Comparison of amino acids: structure 1 is neutral, and structure 2 results from protonation/deprotonation to form the zwitterion.

The overall charge of an amino acid depends on the ratio of negatively charged carboxylates to positively charged amino groups.

The molecule that migrates towards the positive cathode is $H_2NCH_2CHNH_2COOH$ which, at physiological pH, exists as $^+H_3NCH_2CH(^+NH_3)COO^-$.

Therefore, the molecule has a net positive charge.

A: the molecule is neutral with two amino groups protonated and two carboxyl groups deprotonated.

B: the molecule is neutral, with one amino group protonated and one carboxyl group deprotonated. The alcohol has a pK_a of 17 and is neutral at physiological pH.

D: the molecule is neutral, with one amino group protonated and one carboxyl group deprotonated at physiological pH.

6. D is correct.

Free radical halogenation is one of few reactions (along with combustion) that alkanes undergo and occurs via a highly reactive halogen radical.

A radical is a single, neutrally charged atom with an unhybridized *p* orbital with a single unpaired electron that causes it to be reactive.

Steps for free radical halogenation:

> Step I: initiation forms the free radical from a diatomic molecule (X_2) by homolytic bond cleavage. Initiation is shown in the first reaction equation and is usually catalyzed by ultraviolet light ($h\nu$), heat, or by an attack by another free radical.

> Step II: propagation involves a radical and a neutral molecule, and the products are a new neutral molecule and a new radical. The halogen-free radical attacks the neutral alkane and, via homolytic bond cleavage, produces a new H–X bond and another highly reactive free radical as the alkyl radical.

> The alkyl radical reacts with Br_2 to form the alkyl halide and another halogen radical. This new halogen radical then starts the process again, thus causing a chain reaction (propagation).

> Step III: termination involves the joining of two radicals. For example, an alkyl radical attacks a halogen radical to produce a new neutral molecule.

I: $Br_2 + h\nu \rightarrow 2\ Br\cdot$ is the UV light-induced generation of two free radicals and is the chain initiating step.

II: $Br\cdot + RH \rightarrow HBr + R\cdot$ is a chain propagating step because the reaction advances the chain onward by generating a new neutral product plus a new free radical.

III: $R\cdot + Br_2 \rightarrow RBr + Br\cdot$ is a chain propagating step because the reaction advances the chain onward by generating a new neutral product plus a new free radical.

7. B is correct.

Protonation of the alkene results in the formation of the more stabilized carbocation, and this cation forms at the secondary alkyl position.

The cation is then trapped by water to form the alcohol.

8. D is correct.

Alkynes are oxidized by two common mechanisms to yield either the Markovnikov or *anti*-Markovnikov product.

One of the *pi* bonds of the alkyne undergoes addition to yield an enol intermediate.

The enol tautomerizes to generate the Markovnikov ketone or the *anti*-Markovnikov aldehyde.

9. A is correct.

Benzene is aromatic and undergoes electrophilic aromatic substitution (EAS) with the addition of a Lewis acid (e.g., $AlCl_3$, $FeBr_3$ or H_2SO_4).

The reagents of SO_3 and concentrated H_2SO_4 are used to sulfonate aromatic compounds, whereby an SO_3H group is substituted onto the benzene ring.

Halogens are *ortho* and *para*-directing deactivators.

Therefore, the product is a mixture of *ortho*- and *para*-bromobenzenesulfonic acid.

B: the bromine is not displaced from the aromatic benzene ring and replaced with hydrogen to form benzene.

C: the SO_3H group does not substitute for the bromine of bromobenzene.

D: *meta*- is not formed because halogens are *ortho* and *para* directing deactivators.

10. D is correct.

A tosyl group (Tos) is $CH_3C_6H_4SO_2$ (derived from $CH_3C_6H_4SO_2Cl$) and forms esters and amides of tosylic acid.

Tosylates are used to increase the efficiency of the original hydroxyl (~OH) as a leaving group.

Unlike PBr_3 or $SOCl_2$ (both via S_N2), the reaction mechanism preserves the bond between the carbon and the O of the hydroxy, so no inversion of stereochemistry occurs (during the first step in this example) using a tosylate.

The first step with the tosylate results in retention of the chiral center, and the second step (i.e., Cl^- as a nucleophile) produces an inverted product.

11. D is correct.

Since the aldol reaction involves deprotonation (abstraction of H^+) by a strong base, the preferred solvents are neither acidic nor electrophilic.

Dimethyl ether, unlike the other solvents listed, does not contain an acidic proton.

12. C is correct.

Lithium aluminum hydride (LAH or $LiAlH_4$) is a powerful reducing agent and can reduce carboxylic acids and esters to form primary alcohols and reduces nitro groups to amines.

Sodium borohydride ($NaBH_4$) is a weak reducing agent and is only used for the reduction of aldehydes (to primary alcohols) and ketones (to secondary alcohols).

13. B is correct.

Treatment of the ester with potassium hydroxide and heat results in the nucleophilic attack of hydroxide to the carbonyl in an addition-elimination reaction.

The ring product exists as a negatively charged species (cyclohexanol as the alkoxide) because of a base (KOH) until the second step of the H^+ workup.

After acidic workup, the alkoxide is protonated to form cyclohexanol.

14. C is correct.

Amines are Brønsted-Lowry and Lewis bases because the lone pair of electrons on nitrogen can bind to a proton.

It is favorable for the amine to abstract a proton from the acid.

Electron-donating groups attached to the nitrogen in the amine make the amine more basic because by donating electron density, they stabilize the positive ion formed.

Therefore, electron-donating groups (alkyl chains via hyperconjugation) destabilize the lone pair of electrons on the amine and make it more reactive.

Conversely, substituents (e.g., electronegative atoms) are electron-withdrawing, and the amine is less basic.

Compared to hydrogen atoms, alkyl groups are electron-donating, and therefore basicity decreases in the order trimethylamine > methylamine > ammonia in the gas phase.

The gas phase is specified because, in aqueous solutions, hydrogen bonding stabilizes the salt and thus may result in a different order.

In addition reactions, electronegative atoms, such as the fluorine of $(CF_3)_3N$, strongly reduce the basicity of the amine because of the inductive electron-withdrawing effect of the electronegative atom.

In this example, it is unfavorable for the nitrogen to acquire a positive charge in forming a salt.

15. A is correct.

The longest carbon chain has seven carbon atoms; it has one chlorine and one methyl substituent.

IUPAC recognizes the following 5 common names for the nomenclature of organic molecules:

t-butyl   neopentyl   isopropyl

sec-butyl   isobutyl

16. D is correct.

The allylic cation can delocalize the cation at the most substituted position is the most stable molecule.

The other allylic cations are not as stable because the cation is less substituted in the other resonance forms.

17. A is correct.

A racemic mixture contains equal quantities of two enantiomers (i.e., isomers that are non-superimposable mirror images).

Compound I is D-fructose in a Fischer projection.

Compound II is D-fructose in a straight chain.

Compound III is D-glucose in a straight chain.

18. A is correct.

IR absorption between 1630 cm^{-1} and 1740 cm^{-1} is characteristic of carbonyls (e.g., aldehydes, ketones, acid anhydrides, carboxylic acids, esters, and amides).

An IR absorption of 1735 cm^{-1} is characteristic of an ester.

19. C is correct.

For a given class of compounds, the smaller the molecular weight, the lower the boiling point.

The smallest and most volatile components have the lowest boiling point for a mixture containing compounds of different boiling points.

Molecules with the lowest boiling point vaporize first and travel furthest up the fractionating column.

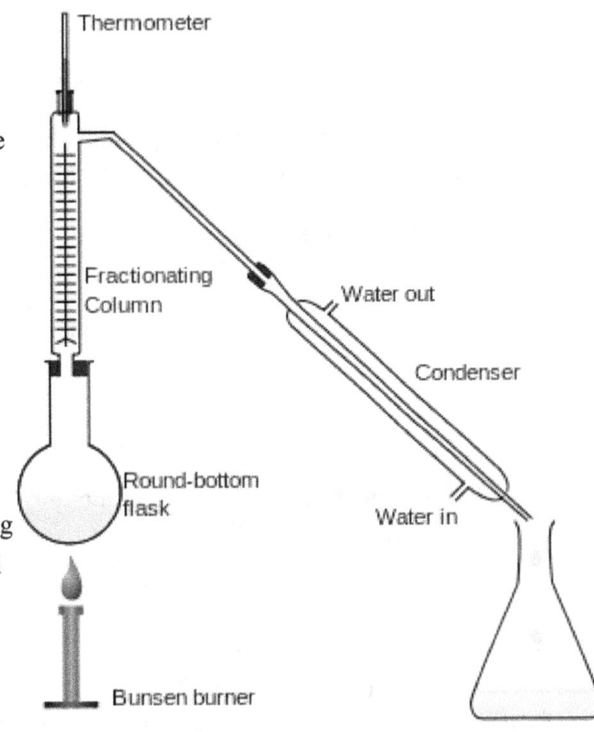

20. B is correct.

As a concerted mechanism, the single step of the S_N2 reaction is a simultaneous substitution occurring with the formation of a new bond while the original bond breaks.

S_N2 undergoes second-order kinetics.

$$\text{rate} = k \, [\text{substrate}] \times [\text{nucleophile}]$$

21. A is correct.

The reaction shown is an acid-catalyzed dehydration reaction. Alcohol in the presence of sulfuric acid and heat is characteristic of the E_1 mechanism to form an alkene.

The generated cation undergoes a ring expansion to give a more stable tertiary carbocation, which is eliminated to form the tertiary (i.e., trisubstituted) alkene.

From the rearranged carbocation, the alpha proton is eliminated by the conjugate base HSO_4^- to regenerate the sulfuric acid.

The ring (i.e., bond angle or Baeyer) strain energy also drives ring expansion.

22. A is correct.

The hydration of the alkyne involves a carbocation and proceeds through a Markovnikov-type mechanism.

The enol intermediate converts (i.e., tautomerizes) to the keto of the methyl phenyl ketone product.

23. A is correct.

Substituents on an aromatic ring affect the rate at which electrophilic aromatic substitution reactions occur.

Since the ring acts as a nucleophile in these reactions, electron-donating substituents increase the reaction rate, while electron-withdrawing substituents decrease the reaction rate.

The bromine substituent (III) is slightly deactivating (due to its electronegativity), making it less reactive than benzene (I).

The nitro group is extremely deactivating (due to resonance), making nitrobenzene (II) less reactive than bromobenzene (III).

24. D is correct.

Aldehydes and ketones tautomerize to exist in equilibrium between the keto and enol forms.

$$R^1R^2\underset{H}{\overset{\alpha}{C}}-\overset{\overset{\textstyle O}{\|}}{C}-R^3 \rightleftharpoons R^1\overset{R^2}{C}=\underset{}{C}\overset{OH}{}R^3$$

Keto   *Enol*

Most molecules (99%) exist predominantly in the keto form because the carbon oxygen (carbonyl) double bond is more stable than the hydroxyl on the double bond of the enol.

Phenols are one of few aldehydes/ketones that exist predominantly in the enol form in the keto-enol tautomer equilibrium. The conjugated benzene ring system of phenol provides stability for the enol form.

The keto form of phenol lacks conjugation because the carbon in the ring is sp^3 hybridized.

When the phenol molecule assumes the keto form, aromaticity is lost, and the molecule becomes less stable.

Enol   *Keto*

Therefore, the keto form is non-aromatic and thus less stable.

Aromatic molecules are cyclic, planar, have conjugated double bonds (i.e., sp^2 at each atom), and satisfy Hückel's number of *pi* electrons ($4n + 2$, where n is an integer).

Anti-aromatic compounds are cyclic, planar, have conjugated double bonds, but have $4n$ (e.g., 4, 8, 12…) *pi* electrons and therefore are unstable.

Nonaromatic compounds do not meet the four criteria needed for aromatic compounds: cyclic, planar, conjugated double bonds, and Hückel's number of *pi* electrons.

25. C is correct.

Oxidation of primary alcohols produces aldehydes (by PCC or oxidation in dry conditions).

Oxidation of secondary alcohols commonly produces ketones (e.g., by PCC) or carboxylic acids (e.g., Jones oxidation by CrO_3 in H_2SO_4).

A: Benedict's test (or Tollens' reagent) detects reducing sugars. An oxidized copper reagent is reduced by a sugar's aldehyde, and the aldehyde is oxidized to a carboxylic acid in the process.

Reducing sugars (and alpha hydroxyl ketones) gives a positive Benedict's test: a red-brown precipitate forms.

Fehling's solution gives a positive test for reducing sugars by changing from blue to clear and forming a red-brown precipitate.

B: Tollens' reagent forms silver ions (i.e., shiny mirror surface) as a positive test for reducing sugars.

26. D is correct.

The chlorine substitution stabilizes the negative charge of the carboxylate through inductive withdrawal of electron density through the *sigma* bonds.

The *ortho* substitution lowers the pK_a of the acid because its conjugation with the aryl ring decreases (*ortho* effect).

27. B is correct.

The longest continuing chain in the product is four carbon atoms long.

The root name is *butan* and the suffix is *~amide* because it is an amide (i.e., R–CONR$_2$) functional group.

For this example, the nitrogen contains a methyl group.

N-methylbutanamide

28. A is correct.

Amines can be protonated with Brønsted acids to produce ammonium salts.

Amides can be formed from primary and secondary amines by acylation of the nitrogen of primary and secondary amines.

29. C is correct.

The longest carbon chain in the molecule is composed of seven carbon atoms.

The methyl substituent is at the fifth carbon of the chain.

The alkene *pi* bond is between carbon one and carbon two.

30. C is correct.

Resonance structures are derived from the movement of lone pairs and *pi* electrons.

Generated negative charges are placed on the more electronegative atoms and the positive charges on the less electronegative atoms.

31. B is correct.

The molecule that contains the chiral carbon is the one that has a central carbon with four different groups as substituents.

These carbons are described as asymmetric (i.e., stereogenic center or chiral carbons).

32. D is correct.

UV light has enough energy to excite electrons to higher energy vacant orbitals to produce photoactivated atoms. This radiation is generally not strong enough to eject electrons from atoms of most elements and form ions.

---

33. A is correct.

Alkanes undergo free radical halogenation to substitute one of the C–H bonds with a C–X bond, where X is a halogen.

Alkenes and alkynes generally undergo addition reactions instead, where an electrophile may add across a carbon-carbon *pi* bond.

34. A is correct.

Both alkanes and alkenes have *sigma* bonds.

The *pi* (i.e., double) bond in the alkenes stabilize a negative charge (as an anion) and, therefore, are more acidic.

The increased *s* character on the hybridization of a sp^2 orbital of an alkene ($pK_a = 45$) allows the orbital to accommodate the negative charge with more stability than the sp^3 of an alkane ($pK_a = 50$).

Furthermore, the increased *s* character on the hybridization of a sp orbital of an alkyne ($pK_a = 28$) allows the orbital to accommodate the negative charge with more stability than for the sp^2 of an alkene

35. C is correct.

The compound has five carbons, seven hydrogens, and one nitrogen.

Use the following formulae to calculate the degrees of unsaturation:

C_nH_{2n+2}: for an alkane (0 degrees of unsaturation)

C_nH_{2n}: for an alkene or a ring (1 degree of unsaturation)

C_nH_{2n-2}: for an alkyne, 2 double bonds, 2 rings or 1 ring, and 1 double bond (2 degrees of unsaturation)

The molecule has 3 degrees of unsaturation.

Because one of the unsaturation elements is a ring, the molecule contains two *pi* bonds.

36. B is correct.

Benzene is a cyclic aromatic hydrocarbon that has 4 degrees of unsaturation.

The *pi* bonds in the molecule are delocalized and impart aromaticity to the molecule.

37. C is correct.

Treatment of salicylic acid (i.e., aspirin) with methanol and dry acid are conditions for synthesizing an ester in a mechanism known as *Fischer esterification*:

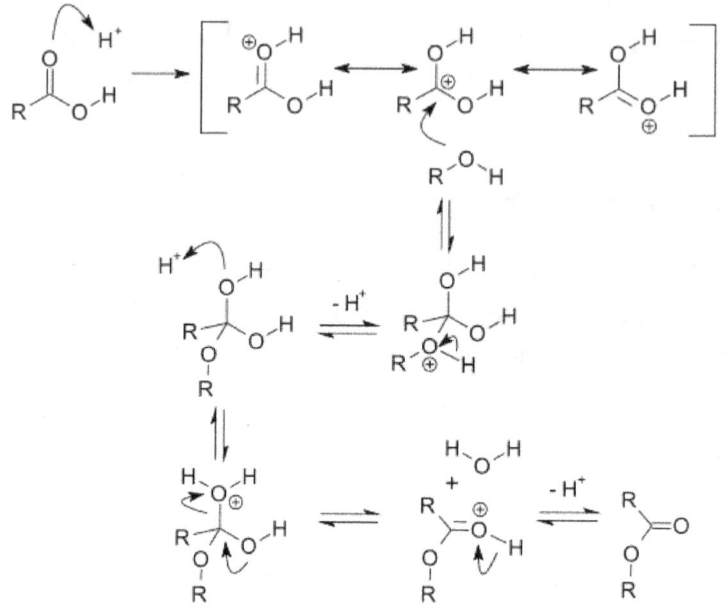

Fischer esterification

38. B is correct.

When ketones and aldehydes are alkylated by Grignard nucleophiles, the number of C–O bonds decreases by one, and the number of C–C bonds increases by one.

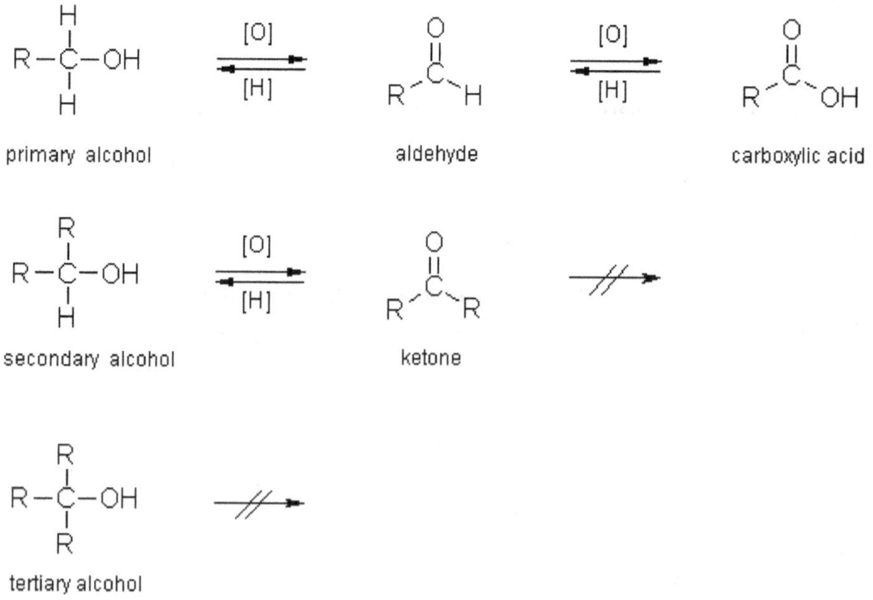

primary alcohol   aldehyde   carboxylic acid

secondary alcohol   ketone

tertiary alcohol

Oxidation proceeds towards the right, while reduction is shown proceeding to the left

39. B is correct.

Exposing a carboxylic acid to sodium hydroxide produces water and the sodium carboxylate conjugate base.

This ionic base has a higher affinity for the aqueous layer due to the negative charge, and this charge forms hydrogen bonds to the hydrogen atoms of water molecules.

40. B is correct.

As the acidity of a group increases, its basic properties decrease.

The carbonyl group of the amide withdraws electron density from the nitrogen atom inductively and through conjugation. This causes the N–H bond of the amide to be much less basic (pK_a of an amine ≈ 10 and pK_a of an amide ≈ 35).

41. D is correct.

Hydrogens, bonded directly to F, O or N, participate in hydrogen bonds.

The hydrogen is partial positive (i.e., delta plus or $\partial+$) due to the bond to these electronegative atoms.

The lone pair of electrons on the F, O or N interacts with the $\partial+$ hydrogen to form a hydrogen bond.

None of the other molecules can form hydrogen bonds because hydrogen is not attached directly to F, O or N.

42. C is correct.

Unless anchored by a high priority group (e.g., carboxylic acid, ketone, or alcohol), the numbering of the longest carbon chain starts at the end that results in the lowest numbering (i.e., the sum of the digits) for the substituent groups.

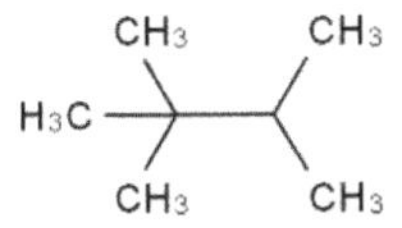

Therefore, the molecule has two methyl groups at the second position and one in the third position: 2,2,3-trimethylbutane.

43. A is correct.

The oxygen atom contains 4 regions of electron density (i.e., two lone pairs and two methyl substituents) and adopts a tetrahedral configuration.

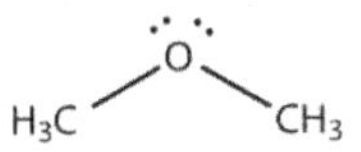

This configuration has an angle between substituent groups to be approximately 109.5 degrees.

44. B is correct.

The chlorine atom occupies the internal (i.e., 2nd) carbon on the first molecule, and the chlorine atom is bonded to a terminal (i.e., 1st) carbon on the second molecule.

Constitutional (i.e., structural or configurational) isomers have the same molecular formula but different connectivity of the atoms.

A: conformational isomers involve free rotation around a single bond (e.g., Newman projections).

C: diastereomers are chiral molecules (i.e., attached to four different groups) with two or more chiral centers and are non-superimposable non-mirror images (i.e., *R,R* and *S,R*).

D: enantiomers are chiral molecules (i.e., attached to four different groups) and are non-superimposable mirror images (i.e., *R* and *S*).

The exception of the two or more chiral center requirements for diastereomers is geometric isomers that contain double bonds (i.e., *cis* and *trans*).

45. D is correct.

The *hydrolysis of esters* using basic conditions (e.g., sodium hydroxide) is not a catalytic reaction. It is a base-promoted reaction because the hydroxide is consumed in the reaction and is not regenerated.

On the contrary, acid-catalyzed hydrolysis reactions of esters regenerate the acid catalyst.

46. C is correct.

Tertiary amines possess no N–H bonds, while secondary amines have one N–H bond, and primary amines contain two N–H bonds.

47. C is correct.

The longest carbon chain is composed of five carbon atoms.

The highest priority group is an aldehyde.

The substituent groups are the hydroxymethyl group, the ethyl group, the alkene, and the alkyne.

48. D is correct.

Trimethylamine $[(\sim NCH_3)_3]$ is similar to ammonia (a central nitrogen atom attached to three hydrogens); trimethylamine has three methyl substituents. The central nitrogen in both examples retains a lone pair of electrons. As lone pairs, the electrons occupy more space than electrons bonded to a substituent.

The negatively charged electrons repel the bonds holding the substituents attached to the central nitrogen atom.

The hydrogens on ammonia have small *van der Walls radii* and are forced closer – bond angle is $107°$.

For trimethylamine, the methyl groups are larger and cannot be forced as close by the lone pair of electrons on the central nitrogen atom. Therefore, the bond angle is approximately $108°$.

49. A is correct.

Isomers are molecules with the same molecular formula but different connections (e.g., functional groups) between the atoms.

Therefore, the number of carbon atoms, hydrogen atoms, and any heteroatoms present must be the same.

50. C is correct.

The stretching frequency is much higher than a typical carbonyl group.

The chlorine atom is a poor electron donor to the carbonyl through lone pair conjugation, and it tends to withdraw more electron density through induction.

The lone pair of electrons on the oxygen atom can donate to carbon-chlorine *sigma* orbital, thus shortening the bond.

The shorter (or stronger) a bond, the higher is its stretching frequency in the IR spectrum.

If the lone pair conjugation with the carbonyl is the more dominant effect (e.g., amides), the carbonyl stretching frequency is lower than 1710 cm^{-1}.

If the inductive effect of the heteroatom outweighs the conjugation effect into the carbonyl group, the carbonyl stretching frequency is higher than 1710 cm^{-1}.

51. B is correct.

S_N1 reactions favor substituted alkyl halides because of the stability of the carbocation intermediates, whereas S_N2 reactions favor unsubstituted reactants due to minimal steric hindrance for the approaching nucleophile.

A: S_N1 reaction rates are greatly affected by electronic factors (degree of substitution and inductive influence of electronegative atoms), while S_N2 reaction rates are greatly affected by steric (hindrance) factors.

C: S_N1 reactions proceed via a *carbocation intermediate*, but S_N2 reactions (by definition as bimolecular) proceed via a single-step reaction involving a concerted mechanism with a transition state instead of a carbocation.

D: S_N2 reactions are bimolecular reactions that proceed via a single-step reaction with a transition state (i.e., bond making and breaking events) rather than via an intermediate.

52. C is correct.

Oxymercuration-reduction of an alkene yields Markovnikov orientation and *anti*-addition, as shown in the cycloalkene below.

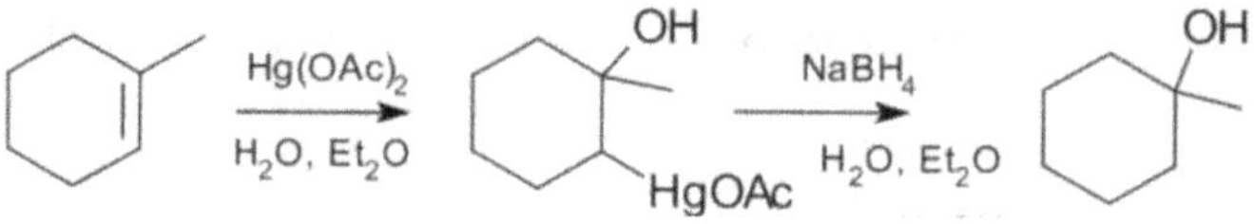

By comparison, hydroboration-oxidation of an alkene is consistent with *anti*-Markovnikov and *syn*-addition.

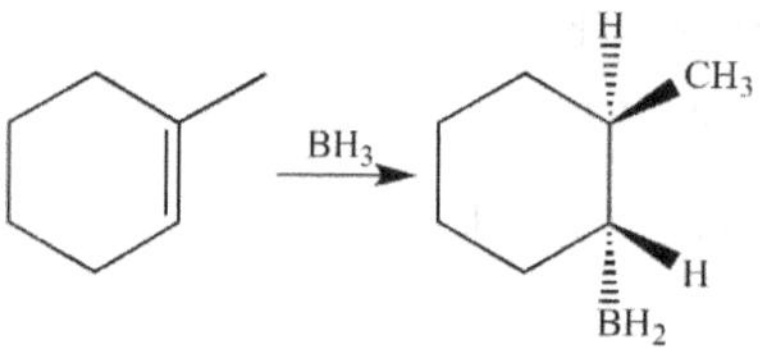

Step one uses BH_3 to add BH_2 and H as *syn* addition across the double bond of the alkene.

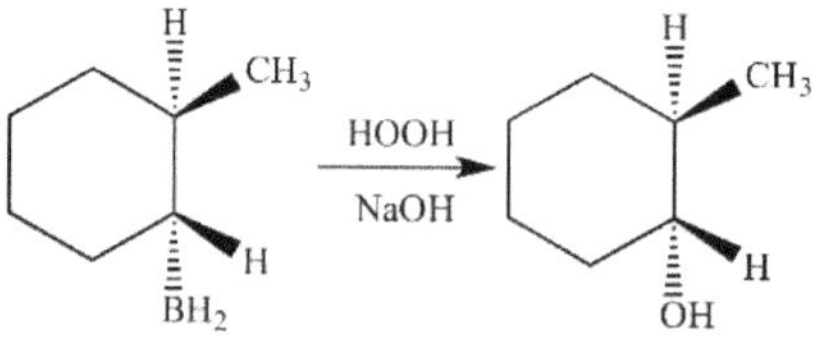

Step two uses peroxides and nucleophilic oxygen for *syn* addition of a hydroxyl.

Other nucleophiles (e.g., methanol) can undergo *syn*-addition as an ether substituent across the double bond.

53. C is correct.

When terminal alkynes are treated with Lewis acids, such as mercury salts in aqueous conditions, the ketone forms as the major product instead of the aldehyde.

The mercury cation is electron-deficient and forms a complex with the alkyne. This coordination increases the electrophilicity of the carbon atoms of the alkyne, and water adds to the internal carbon atom because the partial positive charge is larger at this position. This addition forms an enol (i.e., hydroxyl attached to a carbon in a double bond) that tautomerizes to the ketone.

A reduction step with sodium borohydride ($NaBH_4$) is unnecessary because the carbon-mercury bond could break (to give Hg^{2+} and the enolate) when the ketone forms.

54. D is correct.

The activating, deactivating, and directing (*ortho-* / *para-* and *meta*-directing) properties of aromatic substituents in electrophilic aromatic substitutions (EAS).

Nucleophilic aromatic substitutions (NAS) are based on resonance stabilization or destabilization.

By resonance, electron-withdrawing groups introduce a partial positive charge on positions *ortho* and *para*.

Resonance forms of nitrobenzene during EAS

The *pi* bonds undergo delocalization for two hybrid structures:

In EAS, the aromatic ring acts as a nucleophile, and the partially positive sites are less nucleophilic, making the electron-withdrawing group *meta*-directing. In NAS, the aromatic ring acts as an electrophile, and the partially positive sites are more electrophilic, making the electron-withdrawing group an *ortho-* / *para*-director.

55. A is correct.

The reaction of 1-pentanol + acetic acid

The *alkoxy group* has five carbon atoms, and an acyl portion is an acetyl group.

Acetyl group

Acetate is the root of this molecule.

56. A is correct.

Hemiacetals (shown below) are intermediates produced in acetal formation reactions of aldehydes with two equivalents of alcohols.

Hemiacetals are not stable groups and typically cannot be isolated. The hemiacetal O–H bond collapses to expel alcohol and regenerates the carbonyl of the aldehyde. This typically occurs because the carbonyl formation increases the entropy of the system (starting from one molecule and converting to two molecules), and the carbonyl is almost as thermodynamically stable as the hemiacetal form.

57. D is correct.

The negative charge of the carboxylate anion is stabilized by the electronegativity of the oxygen atom and by resonance stabilization. The negative charge delocalizes over the two oxygen atoms of the carboxylate functional group.

Resonance structures of the carboxylate anion

58. B is correct.

Group 1 is a ketone; 2 is an amide, and 3 is an ester.

Ketones are more electrophilic functional groups than esters.

Esters are more electrophilic than amides.

59. D is correct.

The nitrogen atom of amines is basic because it possesses a lone pair of electrons that add to the acidic protons of acids, resulting in its protonation.

Ammonium cation

The conjugate acid of amines is the ammonium cation, and the nitrogen atom is covalently bonded to four other atoms. Because the nitrogen atom is bonded to four other atoms, it develops a positive formal charge.

Although the nitrogen atom has a formal charge, this positive charge character is distributed over the less electronegative substituent groups.

60. B is correct.

The longest carbon chain has 6 carbon atoms.

There are two methyl substituents in the molecule, and there is an isopropoxide substituent at the C_4 position.

The stereochemistry of the alkene is *E*.

61. A is correct.

Only the hydrogen atom can bond to other atoms with its unhybridized $1s$ orbital.

The *sigma* bonding of other elements is described by hybridizing atomic orbitals before bonding with atoms.

Therefore, when a carbon atom forms a *sigma* bond, it overlaps one of its hybridized orbitals.

However, *pi* bonding involves the indirect overlap of unhybridized *p* orbitals.

62. C is correct.

Cis means that the groups are on the same side of the alkene (or face of a cyclic structure).

Trans means that the groups are on opposite sides.

Cis and *trans* isomerism can occur with disubstituted cycloalkanes.

Cis and *trans* isomerism can occur with alkenes (or with respect to a ring).

63. D is correct.

A: the molecule is *ortho*-fluorobenzoic acid.

B: the substituents should be given the lowest numbering system and alphabetized: 2-chloro-1,3-dinitrobenzene.

C: the substituents should be alphabetized to give 1-bromo-2-iodobenzene.

64. D is correct.

Reasonable resonance forms do not allow nuclei to change positions and satisfy the octet rule. Hydrogen can only have one bond, and carbon can only have four bonds based on valence.

Negative charges should be placed on more electronegative atoms.

65. A is correct.

These molecules are isomers because they have the same number and atoms, but bonding is different.

B: *epimers* are two isomers that differ in configuration at only one stereogenic center. All other stereocenters in epimer molecules, if any, are the same.

C: *anomers* are diastereoisomers for cyclic forms of sugars differing in the anomeric carbon configuration (C–1 atom of an aldose or the C–2 atom of a 2-ketose). The cyclic forms of carbohydrates can exist in either α– or β–, based on the position of the substituent at the anomeric center. α–anomers have the hydroxyl pointing down, while β–anomers have the hydroxyl pointing up.

D: *allotropes* are two or more different physical forms in which an element can exist. Graphite, charcoal, and diamond are allotropes of carbon.

66. C is correct.

3,3-dibromoheptane (below) produces 6 NMR signals.

A: 1,1,2-tribromobutane (below) produces 4 NMR signals.

B: bromobutane (below) produces 4 NMR signals.

D: dibutyl ether (below) produces 4 NMR signals.

67. B is correct.

Fractional distillation separates compounds with boiling points close (differences less than 25 °C).

The extra surface area in the instrument enhances the separation of the compounds.

68. B is correct.

The concentrations of both the substrate and nucleophile control the reaction rate because the rate-determining step is bimolecular.

S_N2 undergoes second-order kinetics whereby:

$$\text{rate} = k\,[\text{substrate}] \times [\text{nucleophile}]$$

69. C is correct.

Conjugation is the alternation of double and single bonds, which results in the delocalization of electrons via resonance through the sp^2 hybridized carbons, resulting in increased stability for the molecule.

1,3-hexadiene has two conjugated double bonds.

A: 1,2-hexadiene has two cumulated, rather than conjugated, π bonds. Cumulated molecules have a sp carbon in the center of the two double bonds – this is unstable and increases the overall energy of the molecule.

B: 1,3,5-heptatriene has three π bonds.

D: 1,5-hexadiene has two isolated π bonds.

70. C is correct.

1-butyne is a gas at room temperature, and 1-propyne has an even lower boiling point, so it is a gas.

Notes for active learning

Notes for active learning

Notes for active learning

Diagnostic Test 3 – Answer Key and Detailed Explanations

#			#		
1	D	Nomenclature	36	D	Aromatic compounds
2	C	Covalent bond	37	C	Alcohols
3	A	Stereochemistry	38	D	Aldehydes & ketones
4	A	Molecular structure & spectra	39	D	Carboxylic acids
5	C	Separations & purifications	40	C	COOH derivatives
6	C	Alkanes & alkyl halides	41	B	Amines
7	D	Alkenes	42	C	Nomenclature
8	A	Alkynes	43	B	Covalent bond
9	D	Aromatic compounds	44	C	Stereochemistry
10	A	Alcohols	45	D	Aromatic compounds
11	D	Aldehydes & ketones	46	A	Alcohols
12	A	Carboxylic acids	47	C	Aldehydes & ketones
13	A	COOH derivatives	48	B	Carboxylic acids
14	C	Amines	49	A	COOH derivatives
15	A	Nomenclature	50	C	Amines
16	D	Covalent bond	51	A	Nomenclature
17	B	Stereochemistry	52	B	Covalent bond
18	D	Molecular structure & spectra	53	D	Stereochemistry
19	B	Separations & purifications	54	C	Molecular structure & spectra
20	A	Alkanes & alkyl halides	55	A	Separations & purifications
21	A	Alkenes	56	D	Alkanes & alkyl halides
22	C	Alkynes	57	A	Alkenes
23	D	Aromatic compounds	58	C	Alkynes
24	A	Alcohols	59	C	Aromatic compounds
25	A	Aldehydes & ketones	60	B	Alcohols
26	B	Carboxylic acids	61	A	Aldehydes & ketones
27	C	COOH derivatives	62	B	Carboxylic acids
28	A	Amines	63	B	COOH derivatives
29	B	Nomenclature	64	B	Amines
30	D	Covalent bond	65	D	Nomenclature
31	C	Stereochemistry	66	B	Covalent bond
32	B	Molecular structure & spectra	67	A	Stereochemistry
33	A	Alkanes & alkyl halides	68	A	Molecular structure & spectra
34	B	Alkenes	69	D	Alkanes & alkyl halides
35	B	Alkynes	70	C	Alkenes

This page is intentionally left blank

1. D is correct.

Draw the four-carbon chain with the double bond at the second position in the chain.

The second and third positions each have a chlorine atom, and these are the highest priority substituents of the alkene. They must be oriented on the same side of the double bond because the molecule is *cis*.

2. C is correct.

The dipole moment is determined by the magnitudes of the individual bond dipoles and the spatial arrangement of the substituents on the molecule.

The dipole moment is greatest when there is a significant difference in electronegativity of the bonded atoms.

A carbon–carbon bond (i.e., same electronegativity) has no dipole moment, while carbon–halogen bonds have moderately large dipole moments because of the electronegativity difference between carbon and halogen.

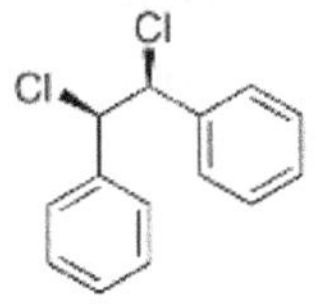

(1*R*,2*S*)-1,2-dichloro-1,2-diphenylethane is effectively *cis* due to the restricted rotation.

(1*R*,2*S*)-1,2-dichloro-1,2-diphenylethane has two phenyl rings attached on one side and two chlorine groups attached on the other side.

The (*R/S*)-designation indicates that the two highest priority substituents (i.e., chlorine) are attached to the same side (priority according to molecular weight, so chlorine has higher priority). The highly electronegative chlorines pull electron density, creating a net dipole.

A: (1*S*,2*S*)-1,2-dichloro-1,2-diphenylethane contains only single bonds. The (*R/S*)-designation indicates that the two highest priority substituents (i.e., chlorine) are attached on opposite sides (with restricted rotation due to the size of the phenyl substitutes). The highly electronegative chlorines pull electron density in different spatial orientations, canceling a net dipole.

B: 1,2-dichlorobutane has carbon–chlorine bonds that are highly polar, but free rotation about the carbon–carbon single bond cancels any net dipole.

D: (*E*)-1,2-dichlorobutene differs from the *Z* configuration (or 1*R*,2*S* in the correct answer) because the two highest priority substituents are on opposite sides of the double bond. As the chlorine pulls electron density, both dipoles cancel for a net dipole of zero.

3. A is correct.

The carbon attached to the leaving group is tertiary (bonded to 3 carbons) and chiral because it is bonded to 4 different substituents, and the molecule is optically active. Tertiary alkyl halides undergo S_N1 reactions (forming a trigonal planar carbocation) but do not undergo S_N2 reactions because of steric hindrance.

In the first step of the S_N1 reaction, the bromine dissociates to form a stable tertiary carbocation, which results in the loss of optical activity. A positively charged carbon is sp^2 hybridized (trigonal planar) and achiral because it has only three substituents.

In the second step, the HCN nucleophile attacks the trigonal planar (flat) carbocation from either side of the plane (i.e., top or bottom) with approximately equal probability. As a result, the reaction yields approximately equal amounts of two chiral products.

The products are enantiomers (i.e., chiral molecules that are non-superimposable mirror images), and each enantiomer rotates the plane of polarized light to the same extent but in opposite directions. Therefore, the product is an optically inactive racemic mixture (i.e., both enantiomers are present in the solution), and there is a loss of optical activity in the solution.

Racemization means loss of optical activity and often involves a carbocation intermediate (S_N1 reaction), whereby the incoming nucleophile attacks from either side of the trigonal planar carbocation.

B: mutarotation occurs in monosaccharides (i.e., sugars) and involves the equilibrium between open-chain forms and cyclic hemiacetal forms (e.g., Haworth projections) in aqueous solutions.

D: inversion of absolute configuration only occurs in S_N2 reactions, whereby a nucleophile attacks the substrate from the side opposite the leaving group (backside) in a one-step reaction. From the concerted S_N2 reaction, the products have the same absolute configuration (often inverted from the backside attack), and the product is considered chiral.

4. A is correct.

UV spectroscopy is useful for identifying compounds that have conjugated double bonds. Neither dimethyl ether nor bromoethane has conjugated double bonds, so UV is not an excellent analytical technique to distinguish them.

B: mass spectrometry (MS) provides information about the molecular weight, the number and size of molecular fragments, and a unique fingerprint pattern.

C: infrared (IR) spectroscopy determines if the molecule contains certain functional groups and gives a unique fingerprint pattern for a molecule.

D: proton nuclear magnetic resonance (NMR) examines the molecular environment of the hydrogens and is related to where the signal is on the spectrum.

NMR is useful in determining the connectivity of the atoms and identifies the relative numbers of each kind of hydrogen (i.e., integration number given by the area under each signal) and the number of hydrogen atoms on adjacent atoms (i.e., splitting pattern as determined by n + 1, where n = number of adjacent H).

5. C is correct.

To separate two compounds, consider their chemical and physical properties. From the question stem, at atmospheric pressure, they both decompose before they reach their boiling point.

Vacuum distillation occurs at very low pressure and causes compounds to boil at lower temperatures – this allows these molecules to reach their boiling point before they decompose.

A: sublimation is the direct conversion of a solid into a gas. This often requires lower pressures and/or higher temperature than vacuum distillation, meaning it is more difficult to carry out and would carry more risk of the compounds decomposing.

B: fractional distillation separates compounds that have boiling points close (differences less than 25 °C). It is usually performed at atmospheric pressure, so (like for simple distillation) these compounds would decompose instead of distilling (evaporating and moving from the liquid to gaseous phase).

D: in simple distillation, a mixture is heated until each component boils (moves from liquid to gaseous phase) separately, and each fraction is collected. Since both molecules decompose before reaching their boiling point at atmospheric pressure, they cannot be separated by simple distillation (they decompose before being separated).

Simple distillation is effective to separate molecules with boiling point differences of more than 25 °C.

6. C is correct.

A *trans* isomer requires that the substituents point in opposite directions (up and down). Therefore, one substituent is axial, while the other substituent is equatorial.

In this substituted cyclohexane, the molecule is more stable when the larger substituents are equatorial.

When comparing methyl and isopropyl, the isopropyl is larger and therefore is equatorial.

7. D is correct.

Carbon-carbon *pi* bonds are elements of unsaturation, and unsaturated compounds can be reduced to give more reduced molecules (e.g., alkanes).

Alkenes have a higher oxidation state than the alkane and are equivalent to a C–O or C–X bond, where X is a more electronegative atom.

8. A is correct.

The Grignard reagent (CH_3CH_2MgBr), as a carbanion, is a strong base.

In the presence of the terminal alkyne shown, an acid-base reaction occurs by deprotonating the alkyne and producing ethane.

9. D is correct.

The double bonds in benzene are less reactive than in a non-aromatic alkene because addition (e.g., hydrogenation) disrupts the aromaticity (i.e., delocalization) of the ring, making it less stable.

The application of heat and high pressure in the presence of the Rh catalyst permits benzene to overcome the energy of activation necessary to transform the highly stable benzene molecule to a non-aromatic product.

10. A is correct.

The electronegativity of the oxygen atom in alcohol helps stabilize the negative charge of the conjugate base (i.e., negative oxygen) of the alcohol.

A more stable conjugate base results in a stronger acid (more readily dissociating its proton).

Secondary alcohols are more acidic than tertiary alcohols.

11. D is correct.

The oxidation of aldehydes to carboxylic acid can be done by exposing aldehydes to chromic acid in water and acetone or to potassium permanganate.

12. A is correct.

Secondary alcohols cannot be oxidized further.

B: primary alcohols undergo oxidation by strong oxidizing agents (e.g., potassium permanganate, O_3 or Jones's reagent) to yield carboxylic acids.

C: acidic or basic hydrolysis of a nitrile yields carboxylic acids.

D: Grignard reagents reacting with CO_2 is a method for preparing carboxylic acids. The carbonation of a Grignard reagent (adding CO_2) forms the magnesium salt of a carboxylic acid. Subsequently, the magnesium salt is protonated and converted to a carboxylic acid when treated with mineral acid (e.g., H^+ from HCl).

13. A is correct.

Consider the susceptibility of different compounds to nucleophilic attack. The molecules undergo nucleophilic attack, but the molecule that undergoes attack the easiest is propionyl bromide.

Out of all the carboxylic acid derivatives (acyl halides, anhydrides, esters, carboxylic acids, and amides), acyl halides are the most reactive towards nucleophiles by the electron-withdrawing effects of oxygen and the stability of the halide anion as a leaving group.

B: benzyl bromide is susceptible to nucleophilic attack because of the electronegative substituents, but not to the same extent as propionyl bromide. Acid halides are more electrophilic than alkyl halides, aldehydes, or ketones.

C: propanal is susceptible to nucleophilic attack since it has a carbonyl carbon but no electronegative substituents, as propionyl bromide.

D: butanoic acid has a carbonyl carbon that is slightly susceptible to nucleophilic attack because the double-bonded oxygen has an electron-withdrawing effect. However, comparing butanoic acid and its functional derivative propionyl bromide, propionyl bromide has withdrawing effects from the oxygen and the bromide.

14. C is correct.

The molecule has a basic site on the nitrogen atom.

The lone pair of electrons on the nitrogen atom can be protonated to form an ammonium cation.

15. A is correct.

The longest continuous carbon chain in this molecule is six atoms long, making it a substituted hexane chain.

The molecule contains a ketone carbonyl (designated by the suffix ~*one*), with the carbon atoms numbered from the end of the chain closest to the carbonyl.

The ethyl substituent is at carbon 3, while the carbonyl is at carbon 2.

Therefore, the IUPAC name for this molecule if 3-ethylhexan-2-one.

16. D is correct.

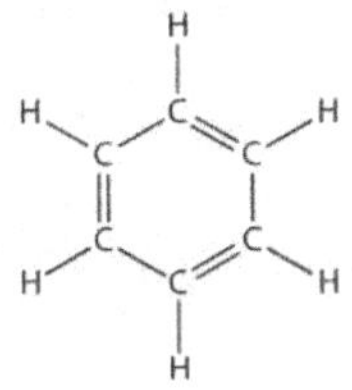

Benzene with *sigma* and *pi* bonds shown

17. B is correct.

Draw the structure of each of the possibilities and count the number of isomers.

Two isomers can be formed from the geminal substitution of the chlorine atoms; three isomers result from the (1,2), (1,3) and (1,4) disubstitution.

The last isomer involves a (2,3) dichloro substitution.

The (2,3) disubstitution can exist as a pair of diastereomers.

18. D is correct.

The compound is an ester.

The 3.8 ppm septet corresponds to the single C–H bond near the oxygen atom of the ester.

The singlet at 2.2 ppm suggests that the CH_3 group is near the carbonyl group.

The doublet at 1.0 ppm corresponds to the methyl groups of the isopropyl portion of the molecule.

19. B is correct.

Water has a ($pK_a \approx 18$).

The nitrogen atom of the amine ($pK_a \approx 10$) can be protonated by the acid to become positive.

The positively charged nitrogen is more soluble in water.

The nitrogen atom of the amine ($pK_a \approx 25\text{-}35$) can be protonated by the acid, but not the water, to become positive.

20. A is correct.

E_1 and S_N1 reactions are strongly favored by highly branched carbon chains and good leaving groups.

E_2 reactions are largely independent of the structure of carbon chains and are favored by good leaving groups, which can easily be eliminated by basic conditions.

S_N2 reactions are strongly favored by substrates with unbranched carbon chains.

$(CH_3CH_2CH_2)_3CBr$ is a tertiary alkyl halide.

B: $CH_3CH_2CH_2CH_3$ is a hydrocarbon where alkanes do not undergo elimination or nucleophilic substitution.

Alkanes are unreactive to most organic chemistry reagents and can either undergo combustion (i.e., burning of propane) or free radical halogenation to introduce a halogen as a leaving group.

C: $(CH_3CH_2)_3COH$ is a highly branched tertiary alcohol carbon chain, so it cannot undergo S_N2.

The ^-OH is an extremely poor leaving group (i.e., unstable anion), so it does not readily undergo substitution.

With heat, alcohols undergo elimination via dehydration (i.e., removal of water).

D: $CH_3CH_2CH_2CH_2Br$ is a primary alkyl halide that undergoes S_N2 and E_2, but neither S_N1 nor E_1.

21. A is correct.

An *alkoxide* is the conjugate base of an alcohol and therefore consists of an organic moiety (i.e., group) bonded to a deprotonated (i.e., negatively charged) oxygen atom.

Secondary halides undergo bimolecular elimination (E_2) with strong bases, especially hindered ones like potassium *tert*-butoxide, $KOC(CH_3)_3$.

B: E_1 designates unimolecular elimination, generally observed in protic (i.e., H^+ donating) solvents (e.g., water or alcohols) and not when subjected to a strong alkoxide base.

C: S_N2 designates bimolecular nucleophilic substitution.

D: S_N1 designates unimolecular nucleophilic substitution.

22. C is correct.

The hydration of the terminal alkyne proceeds with Markovnikov regioselectivity to produce a ketone.

A: the enol intermediate tautomerizes to the keto product.

B: $CH_3CH_2CH_2CH=CHOH$ is the enol intermediate of the *anti*-Markovnikov reaction (hydroboration with BH_3).

D: $CH_3CH_2CH_2CH_2CHO$ is the aldehyde product of the *anti*-Markovnikov reaction (hydroboration with BH_3).

23. D is correct.

$CH_3C_6H_5 + H_2$, Rh/C is a reduction reaction with a powerful reducing agent capable of disrupting the stability of the aromatic ring. The regents reduce the benzene ring catalytically via hydrogenation to form cyclohexane. Therefore, this is not an electrophilic aromatic substitution.

In general, an aromatic ring is especially susceptible to electrophilic aromatic substitution (EAS) in the presence of a Lewis acid (e.g., $FeBr_3$, $AlCl_3$ or H_2SO_4).

A: $CH_3C_6H_5 + C_6H\ CH_2CH_2Cl/AlCl_3$ is an example of Friedel-Crafts alkylation (EAS), whereby toluene (benzene with a methyl substituent) reacts with an alkyl chloride in the presence of the Lewis acid aluminum trichloride ($AlCl_3$). The Lewis acid removes chloride from the alkyl halide, forming a carbocation then attacked by the benzene ring.

B: $CH_3C_6H_5 + Br_2/FeBr_3$ is an example of EAS. $FeBr_3$ (like $AlCl_3$) is a Lewis acid. Toluene is activating, and the Br substitutes in the *ortho/para* position.

C: $CH_3C_6H_5 + CH_3CH_2CH_2COCl/AlCl_3$ is an example of Friedel-Crafts acylation (EAS), whereby toluene reacts with an acyl chloride in the presence of the Lewis acid aluminum trichloride ($AlCl_3$).

The Lewis acid removes chloride from the acyl halide, forming a carbocation which is attacked by the benzene ring. The proton is then removed to restore the aromaticity of the original ring structure.

24. A is correct.

Pyridinium chlorochromate (PCC) is a gentle oxidizing agent which converts primary alcohols to aldehydes.

PCC converts secondary alcohols to ketones.

B: is a carboxylic acid that would require a more powerful oxidizing agent (e.g., Jones reagent; CrO_3, H_2SO_4 and acetone)

C: is an alkene and would proceed via E_1 when the alcohol is subjected to mineral acid (e.g., H_2SO_4)

D: is a terminal alkyl halide produced in two steps. First, the alcohol becomes an alkene (E_1 when the alcohol is subjected to a mineral acid, H_2SO_4).

The alkene is halogenated in *anti*-Markovnikov regiochemistry when peroxides (H_2O_2) are in the reaction.

25. A is correct.

Nucleophiles attack sterically hindered alkyl halides at a much slower rate.

Less sterically hindered substrates undergo reactions with nucleophiles at the fastest rate.

Bromobenzene undergoes the addition reaction at a negligible rate because the ring lacks an electron-withdrawing group that can activate the ring towards nucleophilic aromatic substitution.

Furthermore, the bromide of bromobenzene does not undergo S_N2 displacement reactions because the ring blocks access to the carbon-bromine *sigma** orbital.

26. B is correct.

Acids with smaller alkyl chains have lower boiling points.

Hydrogen bonding increases the boiling point.

Increased molecular mass and hydrogen bonding are factors that increase the boiling point.

Formic acid (below) has the molecular formula of CH_2O_2: it contains one carboxylic acid functional group and a hydrogen atom for the R group.

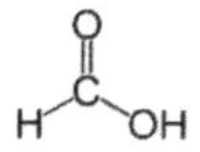

Formic acid

A: oxalic acid (below) has the molecular formula of $C_2H_2O_4$.

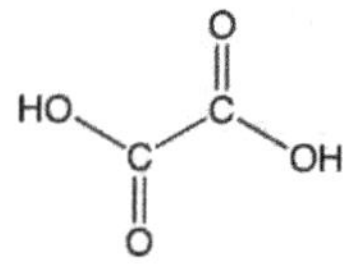

C: benzoic acid (below) has the molecular formula of $C_7H_6O_2$

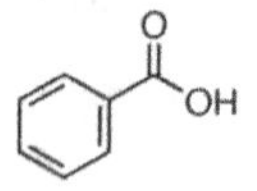

D: acetic acid (below) has the molecular formula of $C_2H_4O_2$.

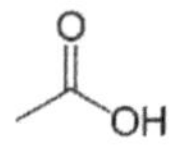

27. C is correct.

Acyl halides and alcohols form esters.

This addition-elimination reaction begins with the nucleophilic attack of the hydroxyl group to the electrophilic carbonyl of the acid halide in an addition reaction.

The elimination of chloride follows to generate the corresponding ester.

Pyridine (below) is a common basic solvent used in organic chemistry:

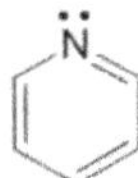

HCl is generated as a byproduct in this reaction; therefore, a base (e.g., pyridine) is needed

28. A is correct.

Hydrogen bonding increases the boiling point of a molecule.

Amines that have N–H bonds typically have higher boiling points compared to tertiary amines.

Tertiary amines can only accept a hydrogen bond, while primary and secondary amines can donate and accept hydrogen bonds.

29. B is correct.

The longest carbon chain is seven carbon atoms and includes an alkene.

The alkene is the highest priority functional group and is assigned the lowest number (i.e., 1 in this example).

Therefore, the molecule has a chlorine substituent in the fourth position.

30. D is correct.

Formal charge = group # – nonbonding electrons – ½ bonding electrons

Nitrogen is in group V on the periodic table.

The ammonium cation has four bonds or eight bonding electrons.

The formal charge for the nitrogen atom is $5 - 0 - 8/2 = +1$.

31. C is correct.

An asymmetric (i.e., chiral) carbon is bonded to four different substituents.

There are three asymmetric carbons in this molecule.

The methylene is symmetrical, the isopropyl group and the geminal dimethyl groups have symmetrical carbons.

32. B is correct.

Conjugated polyenes absorb light at longer wavelengths than unconjugated alkenes because the additional *p* orbital overlap in larger *pi* systems decreases the energy difference between the highest occupied molecular orbital (*HOMO*) and the lowest unoccupied molecular orbital (*LUMO*).

The *LUMO* is an antibonding orbital and has more energy than the *HOMO*, a bonding orbital.

The longer the conjugated system, the smaller the energy gap between the two molecular orbitals (MO); this requires radiation of less energy (and longer wavelength) for electron transitions.

When a molecule absorbs UV/visible radiation, electrons are promoted to a higher energy orbital.

33. A is correct.

In a complete combustion reaction, a compound reacts with an oxidizing element (e.g., oxygen), and the products are compounds of each element in the fuel combined with the oxidizing element.

General formula:

$$C_nH_{2n+2} + [(3n + 1) / 2]O_2 \rightarrow (n + 1)H_2O + nCO_2 + \text{energy}$$

For example, methane yields:

$$CH_4 + 2\,O_2 \rightarrow CO_2 + 2\,H_2O + \text{energy}$$

Nonane:

$$C_9H_{20} + [(3 \times 9 + 1) / 2]O_2 \rightarrow (9 + 1)H_2O + 9\,CO_2 + \text{energy}$$

$$C_9H_{20} + 14\,O_2 \rightarrow 10\,H_2O + 9\,CO_2 + \text{energy}$$

Since nonane has the molecular formula C_9H_{20}, the combustion of 1 mole of neopentane produces 9 moles of CO_2 and 10 moles of H_2O.

34. B is correct.

The bromine adds to the internal position of the epoxide because the partial positive charge is greater at the more substituted position.

The mechanism follows *anti*-addition stereochemistry and would be shown in the product if both the alcohol and halogen were attached to chiral carbons.

35. B is correct.

Alkynes can undergo bromination to yield compounds with four bromine atoms incorporated in their structures.

The first halogenation is expected to proceed more quickly than the second halogenation.

Bromine atoms are large (about the size of a tertbutyl group), and the first bromination increases the steric bulk of the reactant to form the intermediate alkene.

Furthermore, the bromine atoms are more electronegative than carbon, so the *pi* bond of the alkene intermediate is less electron-rich and less nucleophilic than the alkyne *pi* bond.

36. D is correct.

A: of the two substituents, the chloro group is *para*-directing, so it should be substituted first.

Na / NH$_3$ results in the single *trans* hydrogenation of alkenes but does not substitute a nitro group in an EAS reaction.

B: of the two substituents, the chloro group is *para*-directing, so it should be substituted first.

C: while HCl / H$_2$O adds H and Cl across the double bonds of alkenes, these conditions do not substitute Cl in EAS reactions.

37. C is correct.

Tertiary alcohol undergoes E$_1$ reactions faster (i.e., due to the stability of the carbocation intermediate) than secondary alcohols, which can dehydrate at a faster rate than primary alcohols.

38. D is correct.

Benedict's test (or Tollens' reagent) detects reducing sugars. An oxidized copper reagent is reduced by a sugar's aldehyde, and the aldehyde is oxidized to a carboxylic acid in the process.

Reducing sugars (and alpha hydroxyl ketones) give a positive Benedict's test: a red-brown precipitate forms. Fehling's solution gives a positive test for reducing sugars by changing from blue to clear and forming a red-brown precipitate.

The copper complex is reduced to form a red copper product that is less soluble in aqueous solutions.

The detection of this precipitate means the molecule was oxidized.

39. D is correct.

This reaction is the *Fischer esterification* reaction. Water is given off as the byproduct of this transformation.

To drive the reaction equilibrium forward, water should be removed from the reaction, or an excess of one of the two components should be used.

40. C is correct.

This reduction requires two equivalents of lithium aluminum hydride (i.e., powerful reducing agent).

The reactive intermediate involved after the first addition of hydride ($^-$H) is a hemiacetal.

This hemiacetal may reversibly open to form the aldehyde, which can be reduced to a primary alcohol.

Reduction:

carboxylic acid / ester → aldehyde → primary alcohol

ketone → secondary alcohol

Reduction of a carboxylic acid or ester requires $LiAlH_4$.

Reduction of an aldehyde or ketone can proceed with $LiAlH_4$ or the milder reducing agent $NaBH_4$.

41. B is correct.

The most basic site in the molecule is the *N*-methyl tertiary amine because the lone pair of this nitrogen atom is not in conjugation with an electron-withdrawing group or part of an aromatic ring.

42. C is correct.

The longest carbon chain is six carbon atoms.

The two substituents are the chlorine atom and the methyl group.

The highest priority group is chlorine and therefore assumes the lowest number.

43. B is correct.

Pyrrolidine (shown below) is not an aromatic compound, so the lone pair of electrons on nitrogen is available for bonding (i.e., function as a base).

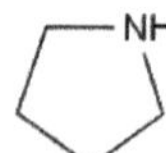

The molecule is a secondary alkyl amine, and the nitrogen atom has sp^3 hybridization.

44. C is correct.

Because one of the stereocenters has a different *R/S* configuration, the molecules are diastereomers.

45. D is correct.

Phenol (i.e., hydroxybenzene) has an ~OH group on the benzene ring.

The hydroxyl oxygen has two nonbonded pairs of electrons, either of which can be donated (via resonance) to the aromatic ring after the addition of an electrophile.

Electron-donating groups stabilize the cations formed upon adding a substituent to the *ortho* and *para* positions and therefore are *ortho/para*-directing activators.

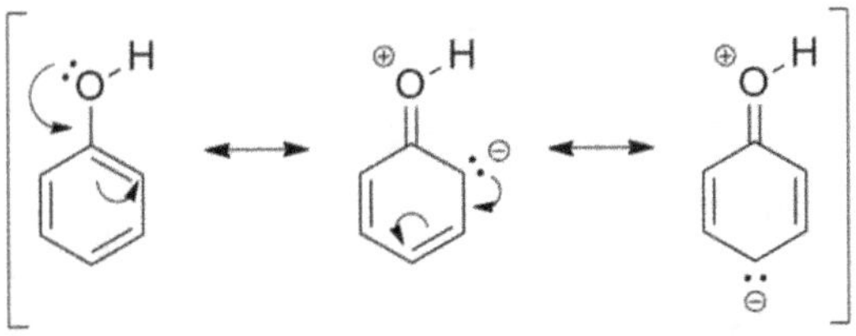

The ~OH group is not notably bulky, and the limited steric hindrance reduces (but not appreciably) substitution at the *ortho* position.

46. A is correct.

The molecule has an alkene on the right side, alcohol near the lower right portion, and a fused ether functional group.

47. C is correct.

Benedict's test detects reducing sugars. An oxidized copper reagent is reduced by a sugar's aldehyde, and the aldehyde is oxidized to a carboxylic acid in the process.

Monosaccharides are reducing sugars, while some disaccharides (e.g., lactose and maltose), oligosaccharides, and polysaccharides are reducing sugars.

Reducing sugars (and alpha hydroxyl ketones) give a positive Benedict's test: a red-brown precipitate forms. Fehling's solution gives a positive test for reducing sugars by changing from blue to clear and forming a red-brown precipitate.

Tollens' reagent forms silver ions (mirror) as a positive test for reducing sugars. The Tollens' test for aldehydes involves the reduction of silver cations [Ag^+] to reduced silver; the metal precipitates out of solution and coats the inner surface of the reaction flask.

Benedict's test, like the Tollens' test, involves the oxidation of an aldehyde to form a carboxylic acid. Sugars able to do this have the aldehyde or hemiacetal functional groups, and these molecules are *reducing sugars*.

48. B is correct.

The carboxylic acid (~COOH) functional group accounts for at least two sites of hydrogen bonding.

Because the answers include carboxylic acids, the hydrocarbon regions of the molecules must be considered.

The molecule with the largest hydrocarbon region has more intermolecular forces (most of which are London forces) and has the largest boiling point.

Stearic acid is a saturated fatty acid and has the highest boiling point.

49. A is correct.

Both acid chlorides and anhydrides combine with alcohols to form esters and with amines to form amides.

Acid chlorides are more reactive than anhydrides, and the chloride is a better leaving group than the carboxylate anion.

These factors contribute to the enhanced reactivity of the acid chloride.

50. C is correct.

Oxygen and nitrogen atoms function as proton acceptors if not conjugated with electron-withdrawing groups.

The amine is the most basic and nucleophilic functional group.

51. A is correct.

The longest chain is the cyclohexane; the chlorines are on the same side (*cis*) and three carbons (1,3) apart.

52. B is correct.

The nitrogen atom does not contain a formal charge (the formal charge calculation for this atom is 0).

However, the charge is slightly negative because the nitrogen atom withdraws electron density from substituent groups via induction, and the lone pair represents a region of electron density.

53. D is correct.

The bromine and hydrogen atoms alternate the carbon atoms they are bonded to, so they are constitutional isomers.

These molecules cannot be chiral because they have a mirror plane of symmetry.

54. C is correct.

Molecules containing the carbonyl ($\sim$C=O) functional groups, such as ketones, typically have C=O resonance frequencies at 1710 cm^{-1}.

However, the stretching frequency is lower if the carbonyl group is in conjugation with another group (in this case, the alkene).

Resonance contributes to the carbon-oxygen single bond character of the carbonyl, and single bonds are weaker than double bonds.

55. A is correct.

Although amines are organic molecules, they have enhanced water solubility in the presence of acid.

The protonation causes nitrogen to develop a positive formal charge, enhancing its dipole interactions with water.

56. D is correct.

The heat of combustion (ΔH_c°) is the energy released (as heat) when a compound undergoes complete combustion with oxygen.

The chemical reaction is typically a hydrocarbon reacting with oxygen to form carbon dioxide, water, and heat. General formula:

$$C_nH_{2n+2} + ((3n+1)/2)O_2 \rightarrow (n+1)H_2O + nCO_2 + \text{energy}$$

The heat of combustion of a compound depends on three main factors: molecular weight, angle strain, and degree of branching.

In most cases, the compound with a higher molecular weight (i.e., more C–C and C–H bonds) has the larger heat of combustion.

For straight-chain alkanes, each addition of methylene groups ($\sim$CH$_2\sim$) adds approximately -157 kcal/mole to the heat of combustion.

For cycloalkanes, the heat of combustion also increases with increasing angle strain.

Reference: 1 kJ$\cdot$mol^{-1} = 0.239 kcal$\cdot$mol^{-1}

ΔH_c° for alkanes increase by about 657 kJ/mol (157 kcal/mol) per $\sim$CH$_2\sim$ group.

For example, heptane has 4 more $\sim$CH$_2\sim$ groups than propane:

$$4 \times -157 \text{ kcal per mole} = -628 \text{ kcal/mol}$$

Yields:

$$-530 \text{ (propane)} + -628 = -1{,}094 \text{ kcal/mol (heptane)}$$

57. A is correct.

Water, methanol, and acetic acid are poor solvents because they can react as nucleophiles and add to alkenes.

58. C is correct.

Two *pi* bonds use one mole of H_2 each, so 2 moles of hydrogen are consumed in the conversion.

59. C is correct.

C: contains 4 π electrons and does not satisfy Hückel's rule for aromaticity ($4n + 2$ π electrons).

Additionally, the silicon atom is sp^3 hybridized and would not participate in aromatic conjugation.

60. B is correct.

Molecules containing rings or multibonds are molecules that have degrees of unsaturation.

A completely saturated compound only consists of single bonds (*sigma* bonds).

61. A is correct.

A common ylide is the phosphonium ylide as a Wittig reagent.

Phosphonium ylide

The double-charged (+ and –) resonance structure is the more dominant resonance contributor because the orbitals of phosphorus are larger than the valence orbitals of carbon, thus making the overlap of the carbon and phosphorus *p* orbitals less favorable.

The minor resonance contributor for this functional group is the charge-neutral pentavalent phosphorus-containing structure formed from the overlap of the C and phosphorus *p* orbitals.

62. B is correct.

Acetic acid is a neutral compound, with the atoms in the molecule having no formal charge.

Carboxylic acids are charged when protonated by other acids (as in Fischer esterification reactions) or deprotonated by bases.

63. B is correct.

The synthesis of amides from carboxylic acids and amines usually requires heating the reaction to high temperatures. This is necessary to create the amide bond and drive away the water byproduct, so the chemical equilibrium favors product formation.

64. B is correct.

Amines are compounds that contain basic nitrogen atoms that can act as nucleophiles if steric interactions do not prevent them from doing so.

The nitrogen lone pair of electrons of the amides are tied up in conjugation with the carbonyl group and are therefore not basic or nucleophilic.

65. D is correct.

The longest carbon chain has five carbon atoms.

The alcohol is attached to the carbon at the second position, and the methyl group is attached to the carbon in the fourth position.

66. B is correct.

Compounds that tend to be polar possess electronegative heteroatoms and more carbon-heteroatom bonds. If these groups can ionize to form charges, then they become even more polar.

67. A is correct.

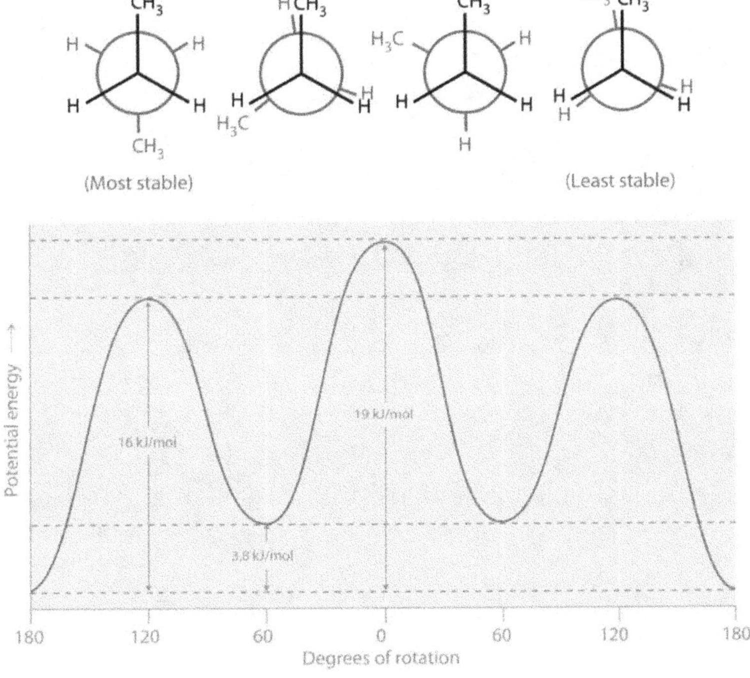

The *anti*-conformer (180° offset) has the lowest energy because it minimizes steric strain (i.e., bulky substituents within van der Waals radii) and torsional strain (i.e., repulsion of bonding electrons when eclipsed).

Gauche refers to a dihedral angle of 60°.

68. A is correct.

The topicity of the protons can be determined by labeling them as H_a and H_b to produce a "chiral center." If this causes the molecule to have two or more chiral centers, then the protons are diastereotopic.

If the labeling causes the molecule to have only one stereocenter, the labeled protons are enantiotopic.

If labeling the protons as H_a and H_b do not lead to the formation of a chiral center (as in methane), then the protons are homotopic.

69. D is correct.

Identify the molecules that contain nitrogen or oxygen heteroatoms because these atoms enable molecules to participate in hydrogen bonding and contribute to dipolar interactions.

Cyclopentane is only composed of hydrogen and carbon, so it is the least water-soluble.

70. C is correct.

The hydration proceeds with Markovnikov addition.

The first step in the reaction is protonation to form the tertiary carbocation; this is trapped by water.

Notes for active learning

Notes for active learning

Diagnostic Test 4 – Answer Key and Detailed Explanations

#	Ans	Topic	#	Ans	Topic
1	B	Nomenclature	36	A	Aromatic compounds
2	A	Covalent bond	37	A	Alcohols
3	C	Stereochemistry	38	B	Aldehydes & ketones
4	A	Molecular structure & spectra	39	A	Carboxylic acids
5	A	Separations & purifications	40	C	COOH derivatives
6	C	Alkanes & alkyl halides	41	D	Amines
7	A	Alkenes	42	B	Nomenclature
8	D	Alkynes	43	A	Covalent bond
9	C	Aromatic compounds	44	A	Stereochemistry
10	D	Alcohols	45	D	Alkynes
11	B	Aldehydes & ketones	46	A	Aromatic compounds
12	C	Carboxylic acids	47	C	Alcohols
13	A	COOH derivatives	48	D	Aldehydes & ketones
14	C	Amines	49	B	Carboxylic acids
15	A	Nomenclature	50	B	COOH derivatives
16	D	Covalent bond	51	A	Amines
17	A	Stereochemistry	52	B	Nomenclature
18	A	Molecular structure & spectra	53	C	Covalent bond
19	C	Separations & purifications	54	D	Stereochemistry
20	A	Alkanes & alkyl halides	55	A	Nomenclature
21	A	Alkenes	56	B	Covalent bond
22	B	Alkynes	57	B	Stereochemistry
23	C	Aromatic compounds	58	D	Molecular structure & spectra
24	B	Alcohols	59	C	Separations & purifications
25	A	Aldehydes & ketones	60	B	Alkanes & alkyl halides
26	B	Carboxylic acids	61	A	Alkenes
27	C	COOH derivatives	62	B	Alkynes
28	B	Amines	63	C	Aromatic compounds
29	C	Nomenclature	64	B	Alcohols
30	B	Covalent bond	65	A	Aldehydes & ketones
31	B	Stereochemistry	66	D	Carboxylic acids
32	D	Molecular structure & spectra	67	C	COOH derivatives
33	D	Alkanes & alkyl halides	68	B	Amines
34	B	Alkenes	69	B	Nomenclature
35	D	Alkynes	70	B	Covalent bond

This page is intentionally left blank

1. B is correct.

The longest chain of carbon atoms is the cyclohexane ring, hence the root of the molecule's name.

There are two methyl substituents at the first and second positions in the ring.

The groups are on the same side of the ring, so they have a *cis* orientation.

2. A is correct.

In carbon–carbon double bonds, there is an overlap of sp^2 orbitals and a p orbital on the adjacent carbon atoms.

The sp^2 orbitals overlap head-to-head as a *sigma* (σ) bond, whereas p orbitals overlap sideways as a *pi* (π) bond.

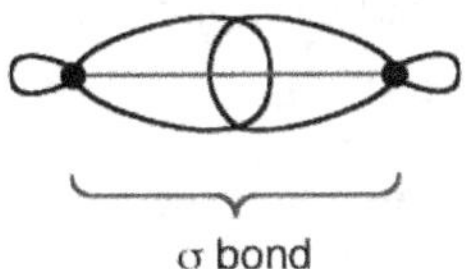

Sigma (σ)bond formation showing electron density along the internuclear axis

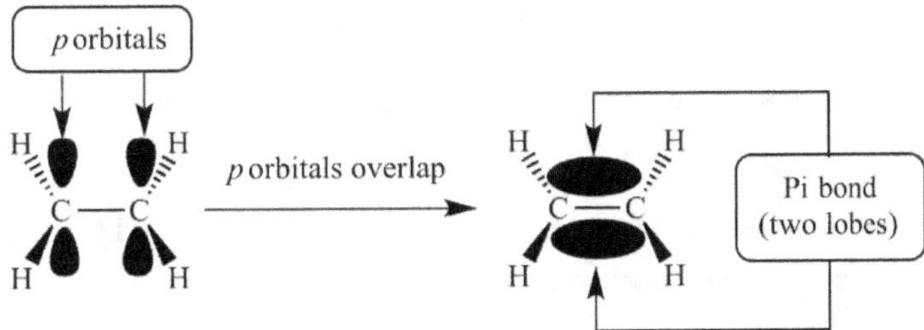

Two *pi* orbitals showing the *pi* bond formation during sideways overlap – note the absence of electron density (i.e., node) along the internuclear axis.

Bond lengths and strengths (σ or π) depend on the size and shape of the atomic orbitals (AO) and the density of these orbitals to overlap effectively.

The σ bonds are stronger than π bonds because head-to-head orbital overlap involves more shared electron density than sideways overlap. The σ bonds formed from two $2s$ orbitals are shorter than those formed from two $2p$ orbitals or two $3s$ orbitals.

Carbon, oxygen, and nitrogen are in the second period ($n = 2$), while sulfur (S), phosphorus (P), and silicon (Si) are in the third period. Therefore, S, P and Si use $3p$ orbitals to form π bonds, while C, N, and O use $2p$ orbitals.

The $3p$ orbitals are much larger than $2p$ orbitals, and therefore there is a reduced probability for an overlap of the $2p$ orbital of C and the $3p$ orbital of S, P, and Si.

B: S, P, and Si can hybridize, but these elements can combine s and p orbitals and (unlike C, O and N) have d orbitals.

C: S, P and Si (in their ground state electron configurations) have partially occupied p orbitals which form bonds.

D: carbon combines with elements below the second row of the periodic table. For example, carbon commonly forms bonds with higher principal quantum number ($n > 2$) halogens (e.g., F, Cl, Br and I).

3. C is correct.

For *Fischer projections*, horizontal lines represent bonds projecting outward (i.e., wedges), whereas vertical lines represent bonds going back (i.e., dashed lines). A Fischer projection does not include a carbon specified at the cross of the vertical and horizontal lines (i.e., a C is implied, but no C is written on the structure).

For assigning *R/S* in Fischer projections, read the ranked ($1 \rightarrow 3$) priorities as clockwise (*R*) or counterclockwise (*S*).

If the lowest priority is vertical (i.e., points into the page), then assign *R/S*. If the lowest priority is horizontal (i.e., points out of the page), reverse ($R \rightarrow S, S \rightarrow R$).

Compound I: the order of priority is hydroxyl, carboxyl, methyl, and hydrogen. The order of increasing priority is counterclockwise, and the configuration appears S.

However, the lowest priority (group 4 is H) is horizontal (pointing outward), so the absolute configuration is *R*.

Compound II: the order of priority is nitrogen, carboxyl, methyl, and hydrogen. The order of increasing priority is counterclockwise.

The lowest priority (group 4 is H) is vertical and therefore points away. The absolute configuration is *S*.

Compound III is achiral because the carbon is not attached to four distinct groups, and therefore the molecule is neither *R* nor *S*.

Compound IV: the order of priority is hydroxyl, carbonyl (aldehyde), methyl (methanol), and hydrogen. The order of increasing priority is counterclockwise.

The lowest priority (group 4 is H) is horizontal and therefore points towards the viewer. The absolute configuration is *R*.

Compounds I and IV have the same absolute configuration.

4. A is correct.

IR active molecules must have polarized covalent bonds to absorb IR. When a Cl–Cl bond with atoms of the same electronegativity stretches or bends, no dipole is created, and therefore the molecule is IR inactive.

B: CO ($C{\equiv}O$) contains covalent bonds whereby carbon is attached to the electronegative oxygen, which creates a dipole generating an IR signal.

C: $CH_3CH_2CH_2OH$ contains covalent bonds attached to an electronegative oxygen, which creates a dipole generating an IR signal.

D: CH_3Br contains covalent bonds attached to electronegative bromines, which create a dipole generating an IR signal.

5. A is correct.

Simple distillation can separate liquids that boil below 150 °C and at least 25 °C apart.

Gas chromatography analyzes volatile (i.e., boiling at low temperatures) liquids.

Recrystallization is commonly used to purify solids, not liquids.

Vacuum distillation is useful when the boiling points are high and close in magnitude. Since there is a difference of 80 °C between the boiling points of the two liquids (i.e., boil at 140 °C and 60 °C), it is not necessary to use vacuum distillation.

Electrophoresis establishes an electrical field and separates molecules based on charge. It analyzes charged species, such as nucleic acids (DNA and RNA) and denatured amino acids (after treatment with SDS, which disrupts hydrophobic interactions).

6. C is correct.

Strong nucleophiles have a negative formal charge (i.e., lone pairs excess), while weak nucleophiles are neutral species with a lone pair of electrons.

7. A is correct.

Product A is an example of *Markovnikov addition*, whereby the hydrogen adds to the least substituted carbon because the most stable carbocation is formed. The bromine then adds to the (most stable) carbocation.

In this example, the hydrogen adds to the secondary carbon, and the bromine adds to the tertiary carbon.

Product B involves free radical intermediates because of the presence of hydrogen peroxide (H_2O_2). Hydrogen peroxide causes the reaction to proceed via a radical intermediate (not carbocation), and the regiochemistry (where the substituents add) is *anti*-Markovnikov. The bromine adds to the least substituted carbon, and the H adds to the most substituted carbon radical.

Product C yields 2-methyl-2-butanol, according to Markovnikov addition.

8. D is correct.

The hydration of the terminal alkyne with BH_3 proceeds with *anti*-Markovnikov regioselectivity.

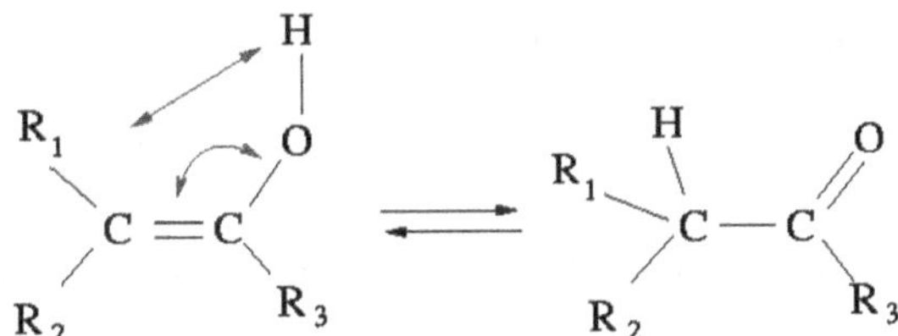

Enol is on the left and the keto on the right

The enol intermediate tautomerizes to the keto product, whereby the keto product (more stable) is over 99% of the observed product.

9. C is correct.

The reactivity of aromatic molecules toward electrophilic aromatic substitution (EAS) depends on the presence of substituents on benzene.

Electron-donating substituents increase the electron density of the benzene ring and therefore activate benzene towards EAS.

Electron-withdrawing substituents deactivate the ring, making it less susceptible to EAS.

The benzene ring is deactivated by the electron-withdrawing effects of the Cl and NH_3^+ substituents, and the ring is deactivated (compared to benzene) to EAS.

A: p-H_3CCH_2O–C_6H_4–O–CH_2CH_3 contains two electron-donating ethoxy substituents and is highly reactive to EAS.

B: p-O_2N–C_6H_4–NH–CH_3 contains strong electron-withdrawing effects from the nitro (NO_2).

The N of the NO_2 group has a formal charge of $+$, while the single-bonded O has a formal charge of $-$.

The NO_2 group offsets the strong electron-donating amino group (lone pair of electrons on N), so the molecule is only slightly reactive to EAS.

D: p-CH_3CH_2–C_6H_4–CH_2CH_3 contains two electron-donating (i.e., activating) ethyl substituents and is more reactive towards EAS.

10. D is correct.

Alcohols of the same chain length as alkanes, alkenes, and alkynes have higher boiling points due to hydrogen bonding of the ~OH group.

Alkanes, alkenes, and alkynes are not able to form hydrogen bonds.

11. B is correct.

Carbonyl groups contain lone pairs on the oxygen atom for hydrogen-accepting capabilities.

However, these groups lack a polarized *sigma* bond to hydrogen atoms (assuming the ketone is in the keto tautomer).

Since these molecules cannot donate hydrogen bonds, they cannot form hydrogen bonds with each other.

12. C is correct.

All the molecules contain the carboxylic acid functional group, which can hydrogen bond and ionize to increase water solubility.

Besides benzene, the choices differ only in the length of the carbon chain attached to the carboxylic acid.

Saturated carbon chains are hydrophobic, and therefore the shortest-chain carboxylic acid is most water-soluble.

13. A is correct.

The amide linkage is present between individual amino acids, and these bonds are commonly known as peptide bonds.

The amide bond is formed when the lone pair of electrons on the nitrogen of the amino group makes a nucleophilic attack on the carbonyl of the other amino acid.

This process is classified as condensation (via dehydration) and results in water loss as the peptide bond forms.

14. C is correct.

The conjugate base of an acid is essentially the deprotonated form.

The most acidic protons of the molecule are the N–H protons, with a pK_a in the mid-30s.

15. A is correct.

The longest carbon chain is 8 carbon atoms.

The only substituent is the isopropyl group.

16. D is correct.

It is helpful to draw the C–H bonds for this compound.

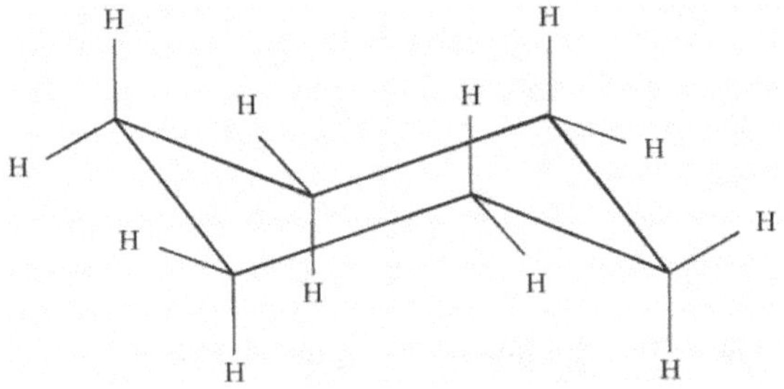

Each carbon atom of the cyclohexane bonds to two hydrogen atoms because there is one degree of unsaturation (the ring).

Six *sigma* bonds exist in the ring, so the number of σ bonds is 18.

17. A is correct.

The root name of the compound is cyclopentane because the ring possesses 5 carbons.

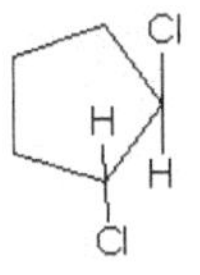

The chlorine substituents are adjacent (i.e., position 1,2) on the ring and must be oriented on opposite sides.

18. A is correct.

Electromagnetic spectrum:

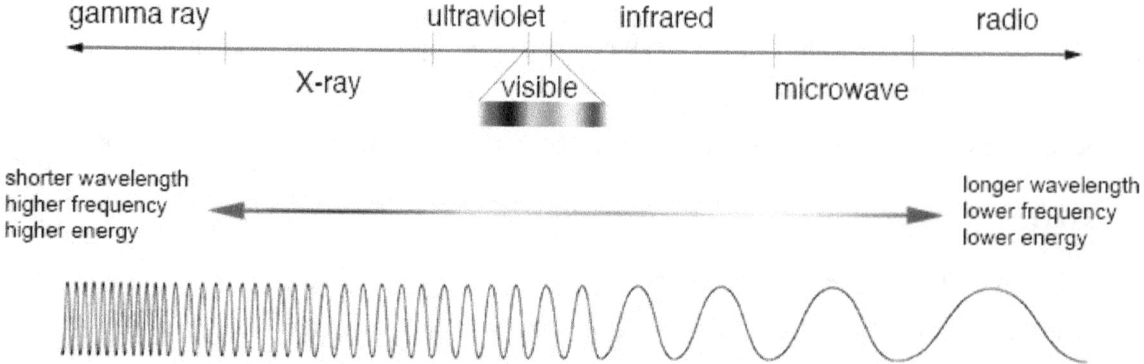

Radio waves have the lowest energy on the electromagnetic spectrum, infrared radiation has more energy than radio waves, and ultraviolet light has more energy than infrared radiation.

19. C is correct.

The free-base form of caffeine is deprotonated and neutral.

Therefore, it has a greater solubility in the nonpolar diethyl ether solvent.

Caffeine (neutral)

The protonated form of caffeine is positively charged and more soluble in the aqueous layer.

Resonance forms of protonated caffeine

20. A is correct.

E_1 refers to unimolecular elimination that proceeds via a carbocation intermediate in a two-step reaction.

Substitution of the substrate ($3° > 2° > 1° >>$ methyl) increases the rate of E_1 (and S_N1) reactions because highly branched carbon chains (with more substituted carbons) form more stable carbocations.

S_N2 refers to the bimolecular nucleophilic substitution that proceeds via a one-step (concerted mechanisms) displacement of a leaving group by a nucleophile.

S_N2 is favored by unbranched carbon chains because the nucleophilic displacement of a leaving group by a nucleophile is favored due to less steric hindrance.

Secondary carbons could undergo both E_1 and S_N2.

The cyanide group is a very poor leaving group because it is an unstable anion and therefore is unlikely to dissociate via E_1 or be displaced via S_N2.

Br^- forms a stable anion (i.e., good leaving group) and dissociates via E_1 or be displaced by S_N2.

B: $(CH_3CH_2CH_2)_3CBr$ is a tertiary alkyl halide and is not favored.

C: $(CH_3CH_2CH_2)_3CCH_2Cl$ is a primary alkyl halide that lacks β-hydrogens and therefore cannot form a double bond by elimination (E_1 or E_2).

D: $(CH_3CH_2CH_2)_2CHCN$ has the leaving group (cyanide) bonded to a secondary carbon.

21. A is correct.

Vinyl refers to an atom attached to carbon on a double bond.

Allylic refers to an atom attached to a carbon adjacent (β) to the double bond.

The chlorine substituent is directly attached to the alkene carbon atoms in vinyl chloride.

22. B is correct.

1-butyne is a terminal alkyne.

Terminal alkynes have additional chemical properties, such as their ability to form anions when exposed to strong bases (e.g., the acetylide anion that forms with $NaNH_2$).

23. C is correct.

The carbon atoms of benzene are sp^2 hybridized.

Non-planar molecules cannot be aromatic because the *pi* system must be planar (i.e., flat).

A Kekule structure is a Lewis structure in which bonded electron pairs in covalent bonds are drawn as lines.

The Kekule structures illustrate the two most significant resonance contributors of benzene.

Benzene with alternating double and single bonds – hybrid structure as the bottom structure

The alternating single and double drawn for benzene exist as a hybrid resonance structure from the delocalization.

24. B is correct.

Carbonyl groups ($\sim C{=}O$) indicate ketones, aldehydes, or carboxylic acid derivatives (acyl halides, anhydrides, esters, and amides).

Ethers are noted as R–O–R and do not contain carbonyl groups.

25. A is correct.

The larger the alkyl portion of an organic molecule, the less likely it is able to dissolve in water.

When ketones and aldehydes are dissolved in water, they are in equilibrium with their hydrated forms.

The hydrated form enhances the solubility of the compound in water.

26. B is correct.

The most polar molecule is the molecule that possesses the smallest alkyl portion.

Acidic functional groups enhance the intermolecular forces between molecules, with hydrogen bonding being the largest contributing factor.

The more polarized a hydrogen-heteroatom bond is, the more acidic it is and the stronger the intermolecular forces the molecule experiences.

Therefore, carboxylic acid is the most polar molecule.

27. C is correct.

The benzene ring (aromatic group) is bonded to the carbonyl of an amide functional group.

28. B is correct.

Amines are one of several functional groups that contain nitrogen atoms.

The other answer choices only contain carbon, hydrogen, or oxygen atoms.

29. C is correct.

When numbering the longest carbon chain of a molecule, start on the end, resulting in the lowest numbering for the substituent groups.

Therefore, the correct molecule should have two methyl groups at the second position and one in the third position.

30. B is correct.

There are four regions of electron density around the nitrogen atom (including the lone pair). Therefore, the nitrogen atom is sp^3 hybridized.

The bonding angles of molecules that possess nonbonding lone pairs of electrons are slightly smaller than predicted by the hybridization state. The nonbonding electrons exert a greater repulsive force than the bonding electrons between the central atom and substituent groups. However, bulky ethyl substituents increase the bond angle from approximately 107° to approximately 109.5°.

Ammonia has a bond angle of approximately 107° due to the electrostatic repulsion of the nitrogen's lone pair on the hydrogen atoms. In amines, as substituents become larger (e.g., $(CH_3CH_2)_3N$), the bond angle between bulky groups increases, and the molecular shape approaches a tetrahedral with a bond angle of 109.5°.

31. B is correct.

Chiral molecules include carbons that are bonded to four different substituents.

This molecule contains no stereogenic centers (i.e., chiral centers), and therefore the molecule cannot be chiral.

32. D is correct.

The type of electromagnetic radiation (EMR) needed to excite an electron in the molecule from the highest energy occupied molecular orbital (*HOMO*) to the lowest energy unoccupied molecular orbital (*LUMO*) corresponds to UV-visible light.

Exciting an electron from the *sigma* bond is more difficult because the *sigma* bond has very low energy and requires higher energy radiation to be promoted to a vacant orbital.

Furthermore, the *sigma** orbital has very high energy; promoting electrons to this vacant orbital requires stronger radiation.

33. D is correct.

Boiling requires the molecules in the liquid phase to overcome the attractive intermolecular forces (e.g., hydrogen bonding, dipole-dipole & London dispersion forces) and move into the gas phase.

The stronger these interactions are, the more energy (i.e., heat) is needed for the molecules to separate from their neighbors and migrate into the gaseous state. Molecules have lower boiling points when branching increases because branching disrupts the spatial packing of molecules in the solid and liquid phase, and therefore branching reduces intermolecular attractions.

A: *cis*-2-pentene has a slightly higher boiling point because unsaturation establishes a dipole moment that raises their relative boiling point compared to alkanes.

B: 2-pentyne has a slightly higher boiling point because unsaturation establishes a dipole moment that raises their relative boiling point compared with alkanes.

C: pentane is a hydrocarbon and only experiences weak London dispersion forces.

34. B is correct.

An *anti*-Markovnikov addition of water across the alkene double bond is needed.

Peroxides (H_2O_2) are a characteristic reagent for *anti*-Markovnikov regiospecificity.

35. D is correct.

The bond order for an alkyne is larger than for an alkene.

The larger the bond order, the shorter the bond.

Therefore, the *pi* bond in an alkyne is shorter.

Furthermore, there is less *p* orbital overlap present in an alkyne than in an alkene.

Because the internuclear overlap is lower, the *pi* bond is weaker.

36. A is correct.

Addition reactions are typically not observed for aromatic compounds because the aromaticity is restored during their substitution reactions.

When aromatic functional groups react, they may temporarily lose their aromaticity (i.e., high-energy resonance hybrids are the intermediates), and its restoration greatly increases the stability of the molecule.

Electrophilic aromatic substitution (EAS) reactions are favored over nucleophilic aromatic addition (NAS) reactions for aromatic compounds.

37. A is correct.

For alkenes, allylic refers to an atom attached to a carbon adjacent (β) to the double bond.

For carbonyl compounds (aldehydes, ketones, acyl halides, anhydrides, carboxylic acids, esters, and amides), the position adjacent to the C of the C=O is the α position.

The hydroxyl group of allylic alcohols is one carbon-carbon *sigma* bond away from the double bond.

Vinyl refers to an atom attached to carbon on a double bond.

38. B is correct.

Grignard + nitrile → imine salt + H_3O → ketone

$$RMgX + R'-C\equiv N \longrightarrow \underset{\substack{\text{imine}\\\text{salt}}}{R-\overset{\overset{\displaystyle NMgX}{\|}}{C}-R'} \xrightarrow[\text{work-up}]{H_3O^+} \underset{\text{ketone}}{R-\overset{\overset{\displaystyle O}{\|}}{C}-R'}$$

Unlike esters, the nitrile is not subject to over-alkylation because the negatively charged imine intermediate generated from the alkylation is less electrophilic than the starting nitrile.

39. A is correct.

Nucleophilic acyl substitution reactions are the common reactions that carboxylic acid derivatives undergo.

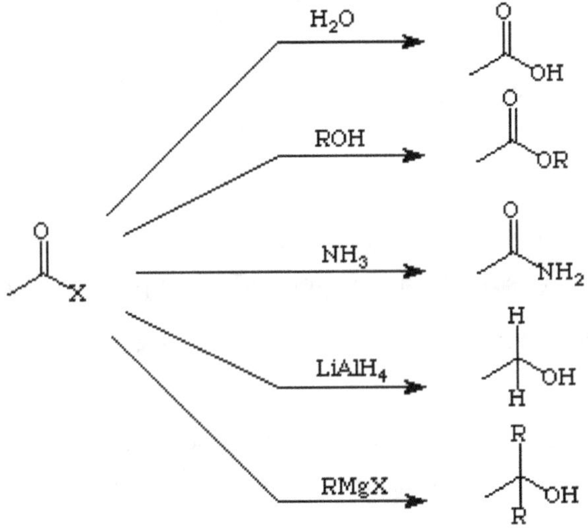

Example reactions involving acyl halides

40. C is correct.

Amides are carboxylic acid derivatives made up of amines, which are carboxylic acid components.

The hydrolysis of carboxylic acid derivatives forms the corresponding carboxylic acid and heteroatom-containing component (e.g., an amine or alcohol).

41. D is correct.

Ammonia groups enhance the water solubility of compounds because of the hydrogen bonding and molecular dipoles that amines possess.

The greater the number of N–H bonds (i.e., primary > secondary > tertiary > quaternary) an amine possesses, the more hydrogen bonds the amine can form and the more soluble it is.

Furthermore, acid can be added to aqueous solutions of amines to enhance the solubility of the molecule.

42. B is correct.

The longest carbon chain in the compound is composed of four carbon atoms.

The highest priority group in the molecule is the amine.

The amine is attached to carbon number 2 (i.e., *sec-* position).

The 5 common names recognized by IUPAC are isopropyl, isobutyl, *sec*-butyl, *tert*-butyl, and neopentyl.

$$H_3C \diagdown \underset{\displaystyle CH_3}{\overset{\displaystyle NH_2}{\diagup}}$$

2-butanamine is the IUPAC name for the molecule

43. A is correct.

Three degenerate *p* orbitals exist for an atom with an electron configuration in the second (n = 2) shell or higher.

The first (n = 1) shell only has an *s* orbital.

The *d* orbitals become available from the third shell (n = 3).

44. A is correct.

The observed rotation is half the value of the specific rotation for the pure enantiomeric substance.

While the effect of the combined opposite enantiomers is canceled, the mixture should have a 50% excess of the pure substance.

Therefore, the mixture must have 75% of the pure enantiomeric substance, where 25% of the rotation cancels the effect of the 25% opposite rotation of its enantiomer.

45. D is correct.

Bromine atoms are added to both sides of the triple bond to produce a vicinal dibrominated alkene.

The addition product has the bromine atoms on opposite sides of the double bond.

Therefore, this addition proceeds through an *anti*-addition mechanism.

46. A is correct.

Four requirements for aromaticity:

    1) molecule is cyclic

    2) molecular is planar (flat)

    3) each atom is sp^2 hybridized (i.e., conjugated)

    4) the number of *pi* electrons satisfies Hückel's rule (4n + 2 *pi* electrons)

Aromatic rings can be positively or negatively charged if the above four criteria are satisfied.

Aromatic compounds have no saturated carbon atoms present in the aromatic ring.

The hybridization state of aromatic carbon atoms is sp^2, and the number of electrons in the aromatic system is consistent with Hückel's rule (4n + 2 π electrons).

47. C is correct.

Primary alcohols have an ~OH group attached to a carbon connected to another carbon atom.

Secondary alcohols possess an ~OH group bonded to a carbon attached to two other carbon atoms.

Tertiary alcohols have ~OH groups bonded to carbon atoms attached to three other carbon atoms.

48. D is correct.

Hydrocarbons tend to have lower boiling points than molecules containing heteroatoms because hydrocarbons lack large molecular dipoles.

These dipoles strengthen the intermolecular forces and raise boiling points.

Furthermore, molecules that can hydrogen bond have the highest boiling points because the hydrogen bond is a strong type of dipole interaction.

49. B is correct.

$$R-\overset{\overset{\textstyle O}{\|}}{C}-OH \; + \; ROH \; \underset{\longleftarrow}{\overset{H^+}{\longrightarrow}} \; R-\overset{\overset{\textstyle O}{\|}}{C}-OR \; + \; H_2O$$

Ester formations from alcohols and carboxylic acids are condensations, so an equivalent of water is lost as a reaction byproduct.

50. B is correct.

6-APA is the core of penicillin and is the chemical compound (+)-6-aminopenicillanic acid.

Since amides are more stable than carboxylic acids, their hydrolysis is higher in energy.

Penicillin with 2 amide bonds

51. A is correct.

When amines are added to water at neutral pH, the solution becomes basic, and hydroxide ions ($^-$OH) are produced.

Alternatively, when added to water at neutral pH, carboxylic acids generate hydronium ions (H_3O^+).

52. B is correct.

The longest carbon chain is a four-carbon cyclo derivative (i.e., cyclopentane).

There are four substituents at the first and third positions in the molecule.

53. C is correct.

No molecular dipole moment exists for symmetric compounds like acetylene because the electron density is evenly distributed on either side of the triple bond.

Asymmetric molecules have molecular dipoles.

54. D is correct.

Draw each isomer and be cognizant of equivalent structures.

ortho-dibromobenzene   *meta*-dibromobenzene   *para*-dibromobenzene

There are only three isomers for the given molecular formula.

55. A is correct.

The longest carbon chain of the molecule is the 3-carbon propane.

This molecule has three substituents: two methyl groups are on the nitrogen atom, and one is at the second position.

56. B is correct.

Tertiary (3°) carbocations are more stable than secondary (2°) and primary benzylic carbocations, which are more stable than primary (1°) or vinylic carbocations.

Methyl cations are the least stable of the carbocations.

57. B is correct.

Isomers are compounds that have the same molecular formula but a different structure.

A: "*hydrocarbons*" refers to organic molecules containing only carbon and hydrogen (i.e., no heteroatoms such as oxygen or nitrogen).

C: *homologs* are a series of compounds with the same general formula, usually varying by a single parameter (e.g., length of the carbon chain).

D: *isotopes* have different mass numbers due to differences in the number of neutrons.

The number of protons determines the identity of the element.

58. D is correct.

The most *deshielded protons* in the NMR spectrum have the largest δ shift (i.e., downfield).

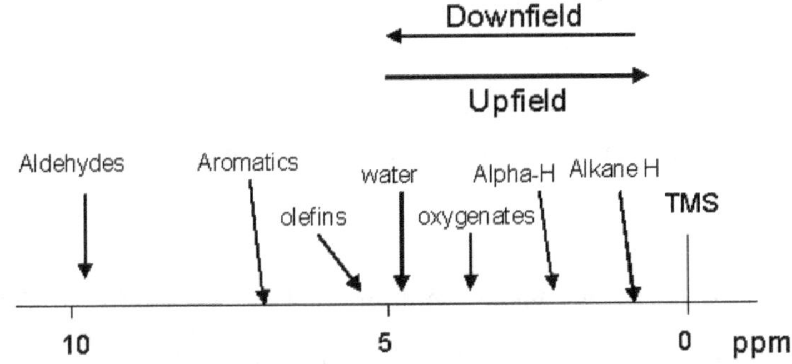

NMR with the approximate δ shifts of some functional groups

59. C is correct.

The *distillation chamber* in the fractionating tower has a constant volume.

Therefore, the pressure is directly proportional to the temperature.

The less volatile components have higher boiling points and therefore require higher temperatures to vaporize them. These high temperatures mean that the pressure will be greater near the bottom of the tower.

60. B is correct.

The order of stability of carbocations is $3° > 2° > 1°$. Therefore, more substituted carbocations are more stable.

61. A is correct.

The rate law may not reveal anything about thel number of steps in a reaction mechanism.

Furthermore, the bromination of alkenes proceeds in two steps, not one.

62. B is correct.

Alkynes are triple bonded molecules made of hydrogen and carbon atoms.

The electronegativity difference between carbon and hydrogen is low, resulting in the molecule being less polar.

Molecules with little polarization may be more soluble in organic solvents as opposed to water.

63. C is correct.

Compared to benzene, electron-donating groups (such as ~OH) activate the ring towards EAS reactions.

Electron-withdrawing groups (such as the acetyl group) deactivate the ring, slowing EAS reactions.

Benzene can undergo the bromination reaction; however, its rate of bromination is less than the rate for the phenol bromination.

64. B is correct.

The ~OH group is characteristic of alcohol and carboxylic acids, but the latter also includes a carbonyl group.

65. A is correct.

The compounds ending in ~*one* suggest that the highest priority functional group is a ketone.

Compounds that end in ~*ol* may possess alcohol as the highest priority group.

66. D is correct.

The acids contain a carboxylic acid functional group, which contributes to the boiling points of molecules.

The molecule with the largest hydrocarbon region has more intermolecular forces (most of which are London forces) and has the largest boiling point.

The alkyl region of stearic acid consists of 17 saturated carbon atoms

Stearic acid is a saturated fatty acid and has the highest boiling point.

Among the answer choices, no other acid possesses an alkyl region this large, and this molecule has the largest boiling point.

67. C is correct.

The reaction above is a *hydrolysis reaction*, and it is the reverse process for the *Fischer esterification* reaction.

In excess water, the acid catalyst activates the ester carbonyl group to promote nucleophilic attack by water.

This produces the corresponding alcohol and carboxylic acid components (in the presence of H^+).

68. B is correct.

The polar amino group of *p*-toluidine makes it slightly soluble in water:

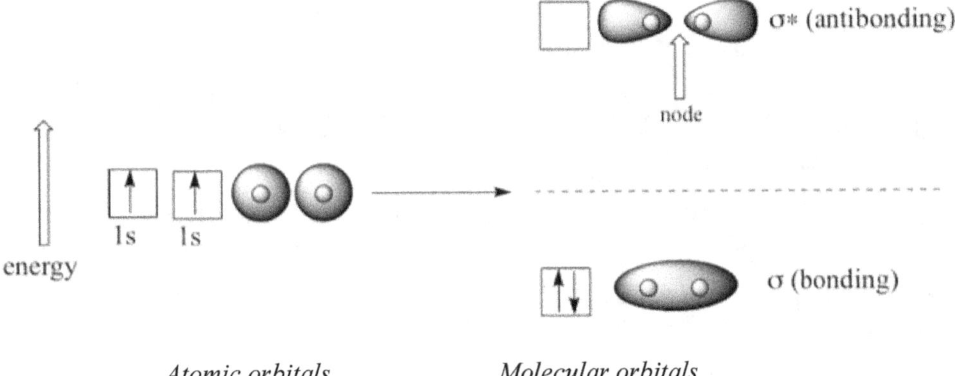

p-toluidine

p-toluidine is soluble in acidic water due to the formation of ammonium salts.

69. B is correct.

The highest priority groups are *trans* to one another.

The longest carbon chain has seven carbon atoms, and the alkene is in the fourth position (fourth carbon from the alcohol, which is the highest priority group).

70. B is correct.

Nodes only form between the orbitals of atoms if the orbital phases have the opposite sign.

Notes for active learning

Notes for active learning

Notes for active learning

Diagnostic Test 5 – Answer Key and Detailed Explanations

#	Ans	Topic	#	Ans	Topic
1	B	Nomenclature	36	D	Aromatic compounds
2	B	Covalent bond	37	C	Alcohols
3	D	Stereochemistry	38	D	Aldehydes & ketones
4	C	Molecular structure & spectra	39	A	Carboxylic acids
5	D	Separations & purifications	40	A	COOH derivatives
6	D	Alkanes & alkyl halides	41	A	Amines
7	A	Alkenes	42	B	Nomenclature
8	C	Alkynes	43	C	Covalent bond
9	D	Aromatic compounds	44	A	Stereochemistry
10	D	Alcohols	45	B	Stereochemistry
11	D	Aldehydes & ketones	46	A	Molecular structure & spectra
12	C	Carboxylic acids	47	D	Separations & purifications
13	C	COOH derivatives	48	D	Alkanes & alkyl halides
14	A	Amines	49	D	Alkenes
15	C	Nomenclature	50	A	Alkynes
16	A	Covalent bond	51	D	Aromatic compounds
17	C	Stereochemistry	52	D	Alcohols
18	A	Molecular structure & spectra	53	D	Aldehydes & ketones
19	D	Separations & purifications	54	D	Carboxylic acids
20	B	Alkanes & alkyl halides	55	A	COOH derivatives
21	C	Alkenes	56	A	Amines
22	D	Alkynes	57	D	Nomenclature
23	C	Aromatic compounds	58	B	Covalent bond
24	D	Alcohols	59	B	Stereochemistry
25	C	Aldehydes & ketones	60	B	Molecular structure & spectra
26	D	Carboxylic acids	61	C	Alkanes & alkyl halides
27	D	COOH derivatives	62	A	Alkenes
28	C	Amines	63	C	Alkynes
29	A	Nomenclature	64	A	Aromatic compounds
30	C	Covalent bond	65	B	Alcohols
31	A	Stereochemistry	66	D	Aldehydes & ketones
32	C	Molecular structure & spectra	67	B	Carboxylic acids
33	C	Alkanes & alkyl halides	68	C	COOH derivatives
34	C	Alkenes	69	A	Amines
35	C	Alkynes	70	A	Nomenclature

This page is intentionally left blank

1. B is correct.

Pentanal:

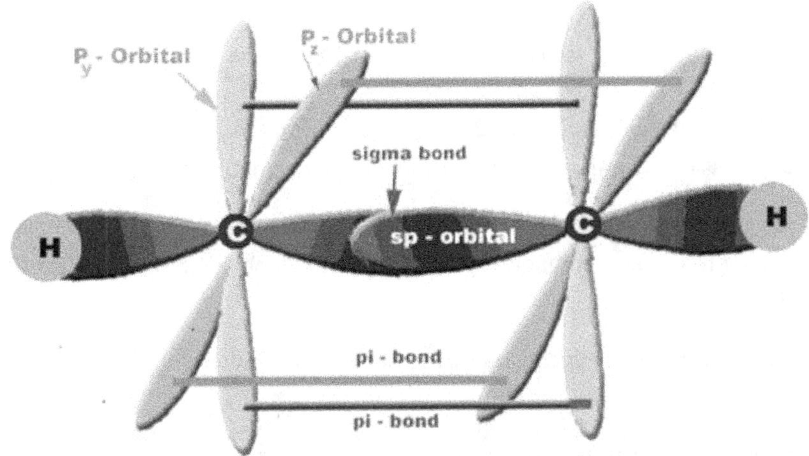

The longest carbon chain is five carbon atoms; the molecule contains an aldehyde, so the suffix is ~*al*.

The suffix ~*one* signifies a ketone, and ~*oic acid* is for carboxylic acid.

2. B is correct.

Acetylene is the common name for ethyne (C_2H_2).

Alkynes are linear and contain a bond angle of 180°.

$$H—C≡C—H$$

Acetylene has a triple bond and therefore contains sp hybridized carbons

A carbon of a triple bond is *sp* hybridized (i.e., *sp* + 2 unhybridized *p* orbitals).

Molecular orbital structure for ethyne showing one sigma bond and two pi bonds

The atoms in a triple bond use *sp* hybridized orbitals from the 2*s* orbital merging (i.e., hybridizing) with a 2*p* orbital.

A: 1,3,5–heptatriene contains sp^2 hybridized orbitals (i.e., double bonds) and sp^3 orbitals for single bonds.

Bond angles of 120° are trigonal planar and originate from sp^2 orbitals. The molecule contains single bonds, sp^3; therefore, a portion of the molecule is tetrahedral.

C: 2-butyne is an alkyne, but only two carbons are *sp* hybridized, while the remaining carbon is sp^3 hybridized, which results from the combination of the 2*s* and the 2*p* orbitals. Four sp^3 carbons form and the bond angle is 109.5° with tetrahedral geometry.

D: dichloromethane has carbon that is sp^3 hybridized because it is attached to two hydrogens and two chlorine atoms. The geometry is tetrahedral for the greatest separation between the substituents, and the bond angle is 109.5°.

3. D is correct.

Structural isomers have the same molecular formula but a different connection of the atoms.

With molecular formula $C_4H_8Cl_2$, the carbon skeleton is butane.

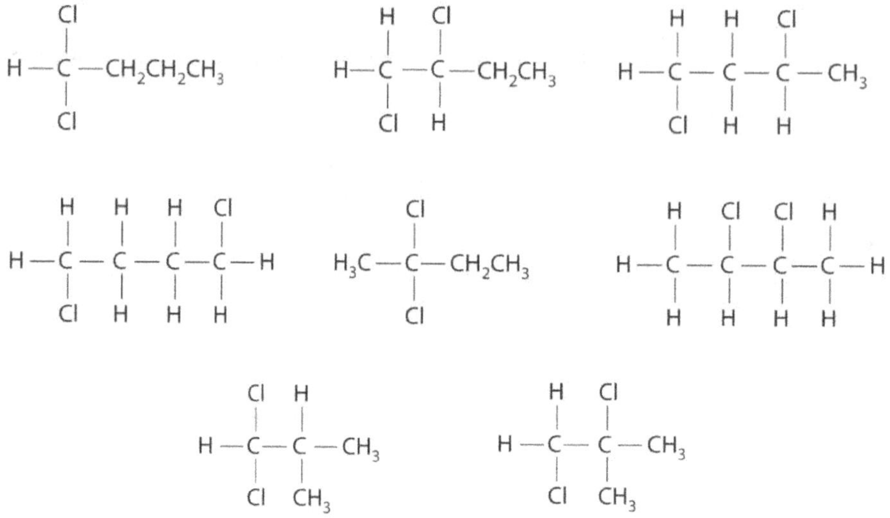

Eight structural isomers exist: three isomers have chiral carbons

From top left and moving to the right: molecules 2, 3 and 6 are chiral (i.e., a carbon attached to four different substituents). Molecule 2 has one chiral carbon; molecule 3 has one chiral carbon.

Structural isomer 6 contains two chiral carbons. One of the stereoisomers of 2,3-dichlorobutane has an internal plane of symmetry, making it the *meso* compound of R/S (or S/R, which is the same molecule) and therefore is achiral and optically inactive.

The other stereoisomer is R/R (or S/S, which is the same molecule) and therefore is chiral and exhibits optical activity.

Therefore, there are three optically active isomers of $C_4H_8Cl_2$.

4. C is correct.

IR spectroscopy provides information about functional groups.

A: mass spectrometry (MS) provides information about molecular weight.

B: nuclear magnetic resonance (NMR) spectroscopy provides information about protons.

D: UV spectroscopy provides information about conjugated (i.e., sp^2 hybridization) double bonds.

5. D is correct.

Higher temperature increases the solubility of most substances because an increase in the kinetic energy allows the solvent molecules to break apart the solute molecules held by intermolecular attractions (e.g., hydrogen bonds, dipole-dipole, and hydrophobic interactions).

The laboratory technique of recrystallization relies upon the principle that a molecule is more soluble at a higher temperature. For example, a compound is more soluble at 40 °C than at 10 °C. Therefore, solubility increases with increasing temperature.

The statement "like dissolves like" refers to compounds that are either polar (or ionic) and dissolve in solvents that are polar (or ionic) or non-polar (hydrophobic) and dissolve in non-polar solvents.

Therefore, molecules and solvents with similar polarities (ionic or hydrophobic) have increased solubility.

The molecular mass of a compound is independent of its ability to dissolve in a given solvent.

6. D is correct.

S_N2 reactions are favored with primary substrates, strong nucleophiles, good leaving groups, and polar aprotic solvents. 1-bromobutane is a primary alkyl halide, ^-CN is an extremely strong nucleophile, and bromine is a good leaving group.

S_N2 occurs exclusively over elimination E_2 because ^-CN is a strong, linear nucleophile; ^-CN is not sterically hindered as a base. Only when strong, bulky bases (e.g., *tert*-butoxide) are used, elimination (e.g., E_2) is favored over substitution (e.g., S_N2).

S_N1 and E_1 reaction mechanisms are not favored with primary alkyl halides (i.e., $\sim CH_2X$) because a primary carbocation is unstable.

7. A is correct.

Hydrogen bromide will not substitute an alkane because alkanes are highly unreactive molecules.

Alkanes undergo substitution under extreme conditions (e.g., *hv* as UV light) or yield combustion products CO_2 and H_2O.

N-bromosuccinimide (NBS) adds bromine to the allylic position (i.e., one away from a double bond). The product is: butene + NBS → 3-bromobutene

Reaction 2: at high temperatures, alkanes undergo combustion to form CO_2 and H_2O.

Reaction 3: free radical substitution is initiated for highly reactive free radicals (e.g., Br_2 or Cl_2), with Br_2 being more selective, whereby the Br radical adds to the more substituted carbon of the alkene.

Reaction 4: Br_2 in CCl_4 adds bromines to an alkene as a bromonium intermediate, with the second bromide adding to the more substituted carbon of the alkene.

8. C is correct.

Use the following formulae to calculate the degrees of unsaturation:

C_nH_{2n+2}: for an alkane (0 degrees of unsaturation)

C_nH_{2n}: for an alkene or a ring (1 degree of unsaturation)

C_nH_{2n-2}: for an alkyne, 2 double bonds, 2 rings or 1 ring and 1 double bond (2 degrees of unsaturation)

Molecular formula $C_{10}H_{16}$ (C_nH_{2n-6}) has 3 degrees of unsaturation. It is consistent with an acyclic molecule that contains two alkyne functional groups or three alkenes or two alkenes and one ring, etc.

A molecule with two triple bonds has 4 degrees of unsaturation (C_nH_{2n-8}) or $C_{10}H_{14}$.

9. D is correct.

Halogens are electron-withdrawing and deactivate a benzene ring toward electrophilic aromatic substitution, so bromobenzene undergoes nitration slower than benzene.

10. D is correct.

Primary alcohols ($\sim CH_2OH$) can be oxidized to the carboxylic acid ($\sim COOH$) functional group with an oxidizing agent (e.g., CrO_3 in HCl).

Oxidation:

primary alcohol $\rightarrow$ aldehyde $\rightarrow$ carboxylic acid

secondary alcohol $\rightarrow$ ketone

tertiary alcohol $\rightarrow$ no reaction

11. D is correct.

The substituents on the carbonyl carbon can be alkyl, alkenyl, or aryl groups.

The R groups do not need to be the same group (i.e., R and R').

Asymmetric ketones form chiral secondary alcohols if reduced or alkylated by a carbon nucleophile that is unlike the structure of the ketone alkyl groups.

Asymmetric ketones are "prochiral" electrophiles.

A: carboxylic acid functional group.

B: ester functional group.

C: aldehyde functional group.

12. C is correct.

The high pH (basic conditions) of the aqueous solution causes the carboxylate to deprotonate and assume its conjugate base form.

This carboxylate form (i.e., an anion of the carboxylic acid) has a higher affinity for the aqueous than the organic layer.

13. C is correct.

The first step in this reaction involves substituting the OH group in benzoic acid with the Cl group from thionyl chloride ($SOCl_2$).

The resulting compound is benzoyl chloride – a highly reactive acyl halide that undergoes nucleophilic substitution. Treating this molecule with NH_3 (ammonia) results in substituting the Cl group with the NH_2 group to form an amide (benzamide).

A and B: benzoyl chloride is highly susceptible to nucleophilic substitution, so chlorine is not part of the final product if another nucleophile (e.g., NH_3) is present.

D: *p*-aminobenzaldehyde is a benzene ring with an ~NH_2 substituent *para*- to the aldehyde group, replacing carboxylic acid. The carboxylic acid would not be reduced to the aldehyde.

The condition for electrophilic aromatic substitution (EAS) requires an electrophile (e.g., Cl_2) and a Lewis acid (e.g., $FeCl_3$).

14. A is correct.

Amines are typically bases due to the lone pair of electrons on the nitrogen.

The name suggests that three methyl groups are bonded to the central nitrogen atom.

15. C is correct.

The three substituent groups attached to the N atom include the two methyl groups and the *tert*-butyl group.

16. A is correct.

Two resonance Kekule structures of benzene

Six *sigma* bonds connect the carbon atoms in benzene.

Furthermore, the delocalized *pi* electron density in the ring is described by three *pi* bonds in resonance.

17. C is correct.

Chiral carbon atoms have four different groups bonded to a carbon atom.

The molecule is named with the alcohol as the highest priority group and designated carbon 1.

The carbon atoms with two or more hydrogen atoms (i.e., carbons 1 and 5) are achiral because at least two groups are the same.

18. A is correct.

The fragmentation pattern of the spectra provides structural information and determination of the molar weight of an unknown compound.

Cleavage occurs at alkyl-substituted carbons reflecting the order observed in carbocations.

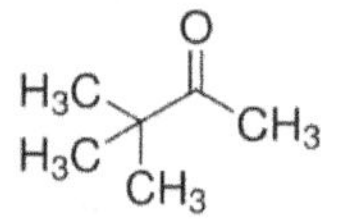

3,3-dimethyl-2-butanone

The base peak for this molecule is the acetyl intermediate. This intermediate results from the ionization of the carbonyl oxygen atom to form an oxygen-centered radical cation.

The carbon-carbon bond between the *tert*-butyl group and the carbonyl can homolytically cleave to give the acylium cation.

For example:

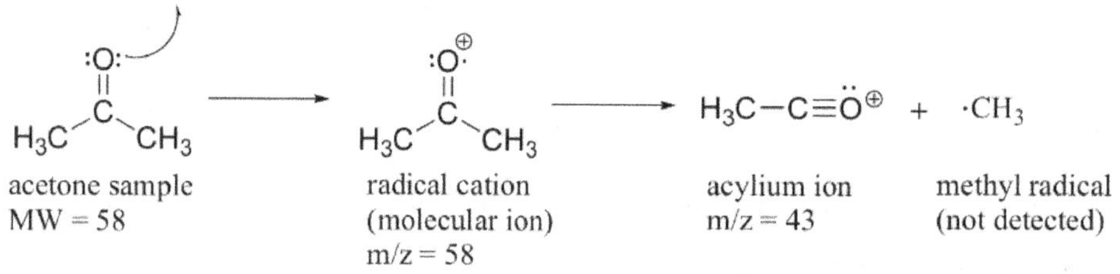

acetone sample
MW = 58

radical cation
(molecular ion)
m/z = 58

acylium ion
m/z = 43

methyl radical
(not detected)

19. D is correct.

Compounds with carboxylic acid ($pK_a \approx 5$) functional groups are soluble in aqueous sodium bicarbonate ($pK_a \approx 10.4$).

The proton of the phenol ($pK_a \approx 9.8\text{-}10$) is not acidic enough to be deprotonated by sodium bicarbonate and impart solubility by creating the alkoxide anion.

20. B is correct.

S_N1 proceeds when the substrate forms a carbocation.

Iodide is the best leaving group (i.e., most stable anion) and forms the cation the fastest.

21. C is correct.

Use the following formulae to calculate the degrees of unsaturation:

C_nH_{2n+2}: for an alkane (0 degrees of unsaturation)

C_nH_{2n}: for an alkene or a ring (1 degree of unsaturation)

C_nH_{2n-2}: for an alkyne, 2 double bonds, 2 rings or 1 ring and 1 double bond (2 degrees of unsaturation)

There are two degrees of unsaturation for the compound C_6H_{10}.

22. D is correct.

A catalytic system, which may produce alkenes from alkynes, is the Lindlar catalyst (i.e., H_2, Pd, $CaCo_3$, quinolone, and hexane).

An alkyne yields a *cis* alkene when subjected to the Lindlar catalyst.

Hydrogenation reactions catalyzed by platinum or palladium result in alkane products.

23. C is correct.

In electrophilic aromatic substitution, the aromatic ring acts as a nucleophile, attacking an electrophile that has been treated with a Lewis acid (e.g., $FeBr_3$, $AlCL_3$).

Deprotonation of the aromatic ring at the site of the attack reforms the double bond (an elimination reaction) and restores aromaticity.

The overall reaction substitutes an electrophile (e.g., Br, CH_3, RCO, HSO_3) for hydrogen on the aromatic ring.

24. D is correct.

The boiling points of compounds are determined by two general factors: molecular weight and intermolecular interactions.

The higher the molecular weight, the harder it is to "push" it into the gas phase, and the higher the boiling point.

Similarly, the stronger the intermolecular interactions, the more energy is required to disrupt them and separate the molecules in the gas phase, hence the higher the boiling point.

Alcohols participate in hydrogen bonding due to the hydroxyl group.

The alkane, alkene, ether, and alkyl halide only participate in dipole-dipole interactions and London forces.

25. C is correct.

A *reducing sugar* can act as a reducing agent because it has a free aldehyde group or a free ketone group.

Monosaccharides are reducing sugars, while some disaccharides, oligosaccharides, and polysaccharides are reducing sugars.

A reducing sugar becomes oxidized (e.g., aldehyde → carboxylic acid) from reducing another compound.

Benedict's test (or Tollens' reagent) detects reducing sugars. An oxidized copper reagent is reduced by a sugar's aldehyde, and the aldehyde is oxidized to a carboxylic acid in the process.

Reducing sugars (and alpha hydroxyl ketones) give a positive Benedict's test: a red-brown precipitate forms.

Fehling's solution gives a positive test for reducing sugars by changing from blue to clear and forming a red-brown precipitate.

The Tollens' test for aldehydes involves the reduction of silver cations [Ag^+] to reduced silver; the metal precipitates out of solution and coats the inner surface of the reaction flask. Tollens' reagent forms silver ions (mirror) as a positive test for reducing sugars. The aldehyde is oxidized to the carboxylic acid when this occurs.

26. D is correct.

The carboxylic acid contains the most acidic functional group among the molecules listed.

Any proton has the potential to protonate a base, given that the base is sufficiently strong to remove the proton from an acid. This requires comparing the pK_a of the acid and base, whereby the base must have a higher pK_a.

Amines can deprotonate carboxylic acids; amide bases can deprotonate alcohols.

The use of strong organometallic bases (e.g., Grignard reagent) may be necessary for the deprotonation of neutral amines and hydrocarbons.

27. D is correct.

The compound contains an aromatic ring called benzyl.

The amide functional group is denoted by:

Where R is an alkyl chain (or H).

An ether functional group is denoted by R–O–R'

Additionally, the molecule contains an aromatic ring, phenol group (i.e., hydroxyl attached directly to a benzene ring), and an alkene (i.e., double bond as a *trans*-alkene).

28. C is correct.

Amine salts are compounds containing a positively charged, tetravalent nitrogen atom and an anionic counterion.

1°, 2° and 3° amines are neutral species.

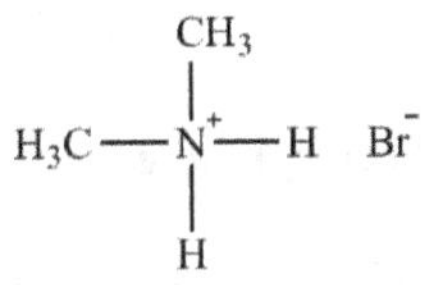

$$CH_3C \equiv N$$

dimethylammonium bromide

A: sulfanilamide (below)

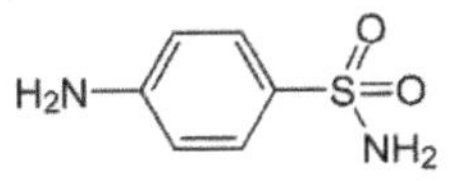

B: thioacetamide (below)

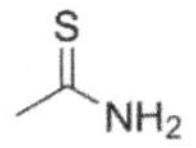

D: histamine (below)

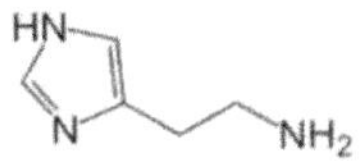

29. A is correct.

The longest carbon chain for this molecule is cyclohexene.

The highest priority group of the molecule is the carboxylic acid (~COOH), and the carbon atom it is bonded to should be labeled as carbon one.

Therefore, the *oxo* (i.e., prefix for the ketone) group is positioned at carbon two.

30. C is correct.

Draw each of the bonds in the structure, where the electrons are distributed to satisfy the octets of the carbon and heteroatoms:

$$CH_3C \equiv N$$

The nitrile has a triple bond composed of a *sigma* bond and two *pi* bonds.

31. A is correct.

Achiral compounds cannot rotate the plane of polarized light.

The solutions of achiral compounds are optically inactive.

B: there is no relationship between absolute configuration (R/S) and the specific rotation (+/–) of light in the polarimeter.

C: *meso* compounds are achiral, but not all achiral molecules are *meso.*

D: *meso* compounds are achiral and contain two or more chiral centers and have an internal plane of symmetry.

32. C is correct.

Topicity is the stereochemical relationship between substituents. These groups, depending on the relationship, can be *heterotopic, homotopic, enantiotopic,* or *diastereotopic.*

The protons are chemically equivalent or homotopic because the groups are equivalent.

If labeling the protons of methylene ($\sim CH_2\sim$) as H_a and H_b do not yield "enantiomers," the molecule is homotopic.

33. C is correct.

Acetate anion

Dimethyl sulfoxide (DMSO) is an organosulfur compound with the formula $(CH_3)_2SO$.

The colorless liquid is an important polar aprotic solvent that dissolves polar and nonpolar compounds and is miscible with water and a wide range of organic solvents.

34. C is correct.

Conjugated (alternating double and single) bonds are more thermodynamically stable than unconjugated double bonds.

B: adjacent double bonds of allenes are not conjugated but cumulated because the *pi* bonds are oriented 90 degrees apart.

Allenes tend to be less stable, especially when confined to cyclic structures.

The other structures are isolated double bonds with (one or more) intervening sp^3 hybridized carbons between the sp^2 carbons of the double bonds.

35. C is correct.

The acetylide anion has a $pK_a \approx 28$.

Due to the significant differences in electronegativity between oxygen and carbon atoms, ions with negatively charged oxygen atoms are relatively more stable than carbon anions.

The sodium methoxide is the most stable conjugate base, and therefore the least basic.

The CH_3Li is the Gilman reagent, and CH_3MgBr is the Grignard. Both are strong bases with a $pK_a >$ than 40.

36. D is correct.

The halogens are deactivating due to their high electronegativity. However, like *ortho / para*-directors, the halogens have a lone pair of electrons on the atom attached to the ring.

Alkyl chains do not have lone pairs of electrons on the C attached to the ring but are *ortho / para* directors due to hyperconjugation.

37. C is correct.

2-hexanol is a secondary alcohol and can be oxidized to a ketone with an oxidizing agent (e.g., CrO_3 in HCl).

The carbon atom of the alcohol is only bonded to one other hydrogen atom; therefore, the highest oxidation state it can acquire is the ketone oxidation state (i.e., +2).

Oxidation: primary alcohol $\rightarrow$ aldehyde $\rightarrow$ carboxylic acid

secondary alcohol $\rightarrow$ ketone

tertiary alcohol $\rightarrow$ no reaction

38. D is correct.

The compounds containing the carbonyl group are ketones, aldehyde, carboxylic acids, and carboxylic acid derivatives (acyl halide, anhydride, ester, and amide).

Other groups may contain carbonyl groups, such as carbonate, carbamate, urea, etc., and these groups have the same oxidation state as carbon dioxide with four C–X bonds.

39. A is correct.

Citric acid is a chemical involved in the citric acid (TCA or Krebs) cycle.

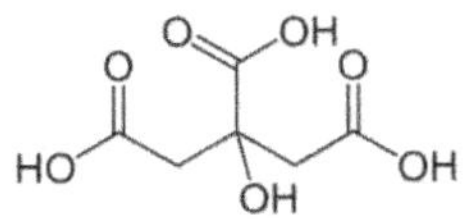

Citric acid

The TCA (Krebs) cycle is a metabolic process responsible for ATP production and commonly occurs in most aerobic organisms.

40. A is correct.

Fischer esterification involves a carboxylic acid and alcohol with an acid catalyst.

Ethyl propanoate

Esters have the general formula: R–COO–R'

The suffix for an ester is –*oate*. The prefix is the substituent attached to the oxygen adjacent to the carbonyl carbon, and the root is the substituent attached to the carbonyl oxygen.

Fischer esterification

Excess alcohol drives the chemical equilibrium forward.

41. A is correct.

Amines tend to be basic and nucleophilic if they are not sterically bulky.

The attachment of acyl groups to amines causes the nitrogen lone pair to delocalize into the carbonyl π^* orbital, thus greatly reducing the nitrogen's nucleophilic and basic properties.

The amide ($RCONR_2$) group reacts with electrophiles on the carbonyl oxygen atom because a more stable cationic intermediate is generated.

42. B is correct.

The longest carbon chain is composed of seven carbon atoms.

The remaining carbon groups are substituent methyl groups at the second, fourth, and fifth positions.

43. C is correct.

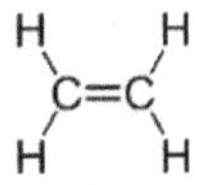

ethene

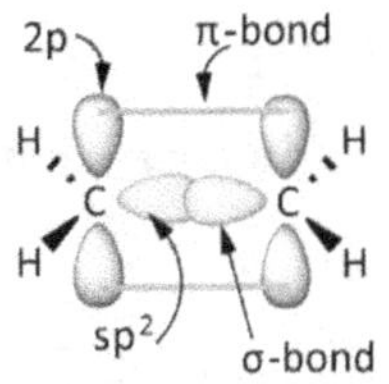

Hybridization and sigma and pi bonds indicated

Hybridization	Bond angle	Geometry
sp^3	109.5°	tetrahedral
$sp^2 + p$ (unhybridized)	120°	trigonal planar (flat)
$sp + p + p$ (two unhybridized)	180°	linear

44. A is correct.

The stereochemistry about the alkene is '*E*' (highest priority groups are on opposite sides of the alkene).

The priority groups are ranked according to the Cahn-Ingold-Prelog rules for prioritization based on the atomic number of atoms attached to the alkene.

The chain possessing the heaviest atoms proximal to the alkene generally has higher priority. In this example, the bromomethyl group has a higher priority.

The alcohol and methoxy groups contain oxygen atoms, but the methoxy oxygen atom is closer to the double bond.

45. B is correct.

(Z)-1-bromo-1-chloropropene   *(E)-1-bromo-1-chloropropene*

These molecules can only adopt one conformation because of the rotational barrier of the alkene.

Configurational isomers involve a double bond.

If the molecules have optical activity, they may be enantiomers (non-superimposable mirror images) or diastereomers.

If they lack optical activity, they are geometric isomers. (see diagram below)

A: *constitutional* isomers have the same molecular formula but different connectivity.

The molecules are not the same, but the connectivity is the same, so they are not constitutional isomers.

C: identical molecules may be drawn with different orientations on the paper but are the same.

D: *conformational* isomers refer to different molecules due to free rotation around single bonds (e.g., Newman projections or chair flips for cyclohexane).

See schematic of isomers below.

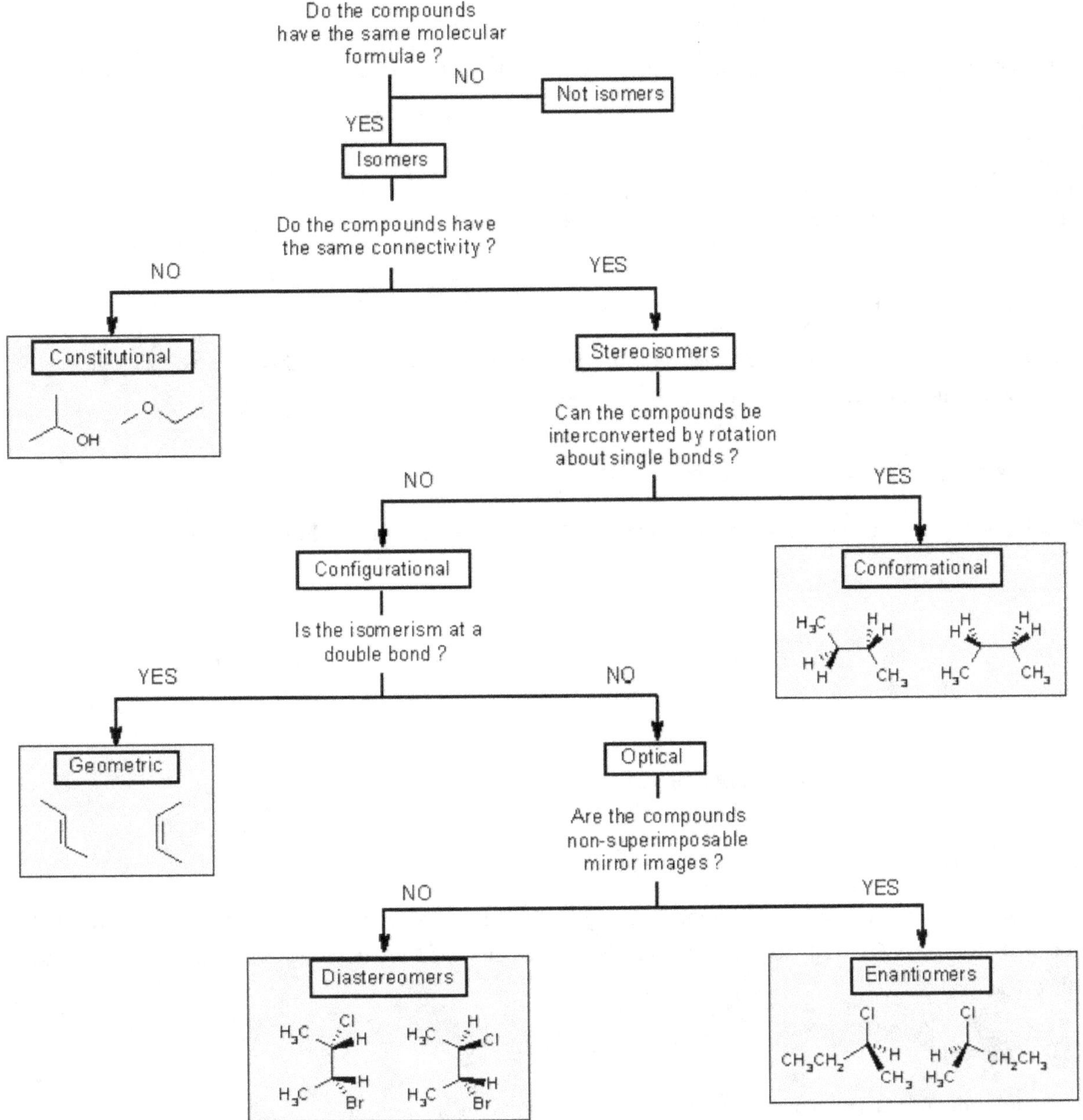

46. A is correct.

The M–18 peak corresponds to the loss of water. The loss of water in a substrate containing alcohol may suggest that an alkene intermediate is generated during the ionization process.

47. D is correct.

Isomers are compounds that have the same molecular formula but a different structure.

Isomers may or may not have the same physical properties. For compounds that are not isomers of each other, different physical properties should be expected.

Enantiomers are the only types of isomers that have the same physical properties. Enantiomers require a chiral environment (e.g., a chiral substrate such as an enzyme in the body) for their separation.

48. D is correct.

1-bromopropane is a primary halogen and a strong nucleophile, leading to S_N2, preferred when there is no steric hindrance.

49. D is correct.

Draw the structures of the starting material and product to determine which reagent is used.

Two halogen atoms are added to the molecule, and an efficient method to achieve this is by exposing the alkene to a diatomic halogen species (e.g., Cl_2 or Br_2).

The chlorine atoms add in an *anti*-orientation from the 3-membered ring of the chloronium ion.

bromoniom ion

Bromonium (like chloronium) ions undergo the same reaction mechanism

50. A is correct.

Combustion is when a hydrocarbon is burned (adding O_2), the major byproducts of the reaction are water (H_2O) and carbon dioxide (CO_2). Ash may form from the reaction, composed of carbon material that cannot undergo further oxidation.

51. D is correct.

The *pi* molecular orbitals of benzene are made from the overlap of six *p* orbitals.

Because the number of molecular orbitals (MO) equals the number of atomic orbitals (AO) involved in the overlap, there are six molecular orbitals.

52. D is correct.

Sodium dichromate ($Na_2Cr_2O_7$) is a strong oxidizing agent which converts primary alcohols to carboxylic acids.

Sodium dichromate converts secondary alcohols to ketones.

53. D is correct.

These reaction conditions are the *Clemmensen reduction* reaction conditions.

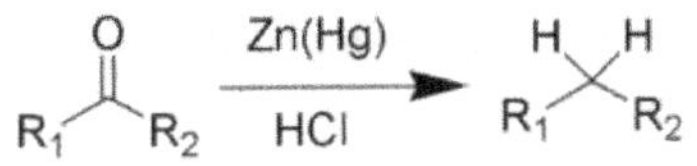

The Clemmensen reduction mechanism involves the protonation of the ketone oxygen atom.

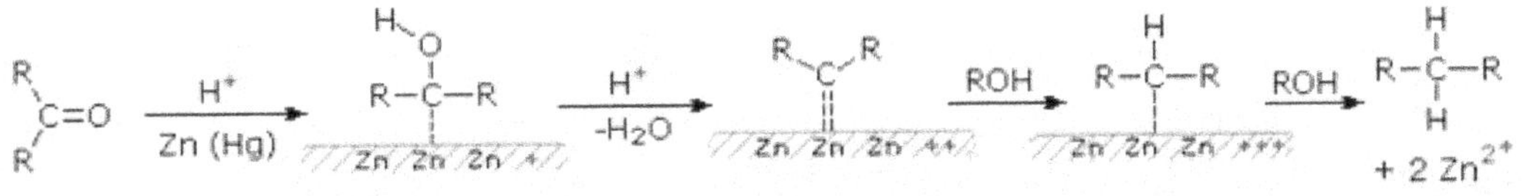

Clemmensen reduction reaction

54. D is correct.

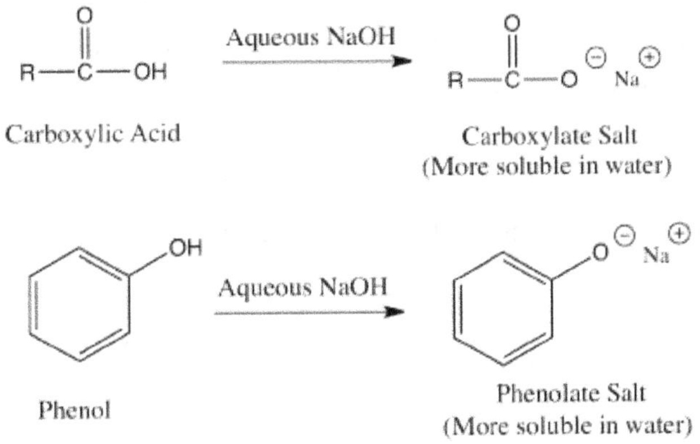

Carboxylic Acid → Carboxylate Salt
(More soluble in water)

Phenol → Phenolate Salt
(More soluble in water)

The salt is a sodium carboxylate salt of the carboxylic acid, and the byproduct in the reaction is water.

Because carboxylic acids are about 10 or more orders of magnitude more acidic than water, the conversion to the carboxylate form by exposing it to sodium hydroxide is irreversible.

Therefore, no equilibrium is associated with this reaction.

55. A is correct.

Sodium hydroxide can act as a base to reversibly deprotonate the N–H of the amide (due to resonance stabilization of the anion).

However, the irreversible reaction pathway forms carboxylate and amine products.

56. A is correct.

Amines are derivatives of ammonia. These compounds contain carbon-nitrogen single bonds and are known to be sufficient bases for strong acids.

Carbonyl-containing (C=O) compounds are aldehyde, ketone, acid chloride, anhydride, carboxylic acid, ester, and amide.

57. D is correct.

The longest carbon chain contains five carbons.

The highest priority group in the molecule is the alcohol, so the compound has *~ol* as the suffix and *pent~* as the root name.

The methyl substituent is in the second position.

58. B is correct.

As the number of covalent bonds (i.e., triple > double > single) increases between carbon atoms, the bond order and bond strength increases.

Stronger bonds are shorter, so the two carbon atoms that make up the triple bond have the shortest bond.

59. B is correct.

Chiral carbons refer to asymmetric carbons that are typically bonded to three other atoms.

Carbon 2 cannot be an asymmetric carbon (or chiral carbon) because carbon atoms with trigonal planar configuration possess a plane of symmetry and are achiral.

60. B is correct.

The *alpha nitrogen* atom makes the neighboring C–H bonds less shielded because the nitrogen atom is electronegative and withdraws electron density through an inductive effect.

Terminal or substituted alkyl C–H bonds (away from electronegative atoms) resonate between 1 and 2 ppm.

61. C is correct.

Molecules with double or triple bonds cannot undergo free rotation because the *pi* bond rigidifies the structure of the compound. To achieve the bond rotation, the *pi* bond(s) must be broken.

Furthermore, for cyclic compounds, the smaller the ring size, the fewer degrees of freedom it has.

Due to strain energy, cyclopropane is unable to rotate freely.

62. A is correct.

The *Zaitsev product* is the most thermodynamically stable.

Thermodynamically stable alkenes are most substituted ($3° > 2° > 1°$).

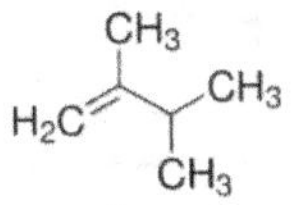

2,3-dimethyl-1-butene has a 1° alkene carbon (terminal) and a 3° carbon (at position 2)

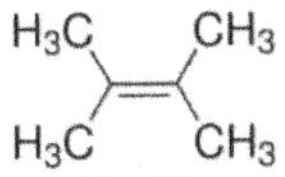

2,3-dimethyl-2-butene has two 3° alkene carbons at positions 2 and 3

The hyperconjugation of the neighboring C–H bonds into the pi^* orbital of the alkene lowers the energy of the alkene.

Furthermore, the sp^3 hybridized alkyl groups inductively donate electron density to the more electronegative sp^2 hybridized carbon atoms.

63. C is correct.

Alkyne: C_nH_{2n-2} = 2 degrees of unsaturation due to the triple bond.

In the name of this molecule, the prefix is *pro–*, which indicates that the molecule is composed of three carbon atoms.

The *~yne* suffix indicates that a carbon-carbon triple bond exists in the molecule.

Therefore, propyne has three carbon atoms and four hydrogen atoms.

64. A is correct.

Four requirements for aromaticity:

    1) molecule is cyclic

    2) molecular is planar (flat)

    3) each atom is sp^2 hybridized (i.e., conjugated)

    4) the number of *pi* electrons satisfies Hückel's rule (4n + 2 *pi* electrons)

Aromatic rings can be positively or negatively charged if the 4 criteria are satisfied.

B, C, and D are *antiaromatic* with 4n *pi* electrons.

65. B is correct.

Thionyl chloride ($SOCl_2$) is a reagent that converts primary and secondary alcohols to alkyl chlorides.

Phosphorous tribromide (PBr_3) is a reagent that converts primary and secondary alcohols to alkyl bromides.

66. D is correct.

When Grignard (R–MgBr or R–MgCl) or Gilman (R$_2$CuLi) reagents are combined with aldehydes, the product is a secondary alcohol.

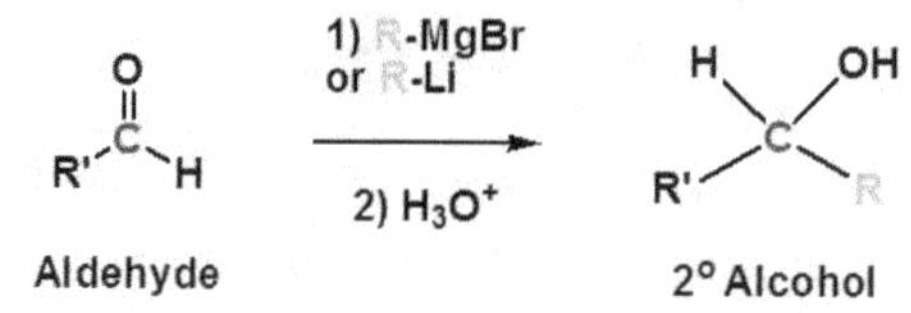

This alcohol may or may not be chiral, depending on which Grignard reagent is used for alkylation.

The alkylation of ketones yields tertiary alcohols.

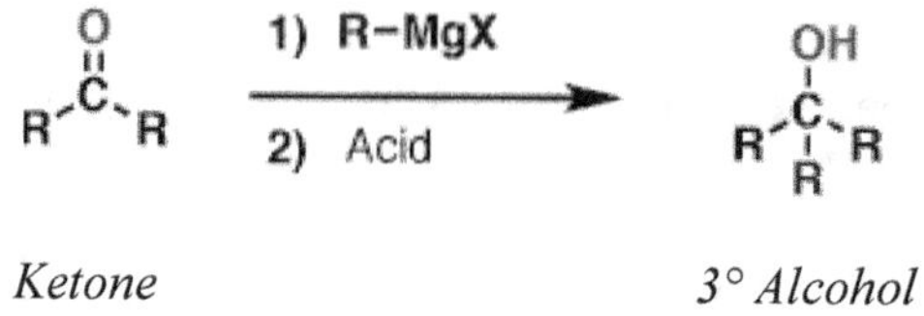

The alkylation of epoxides (below) results in an alcohol with an extended carbon chain.

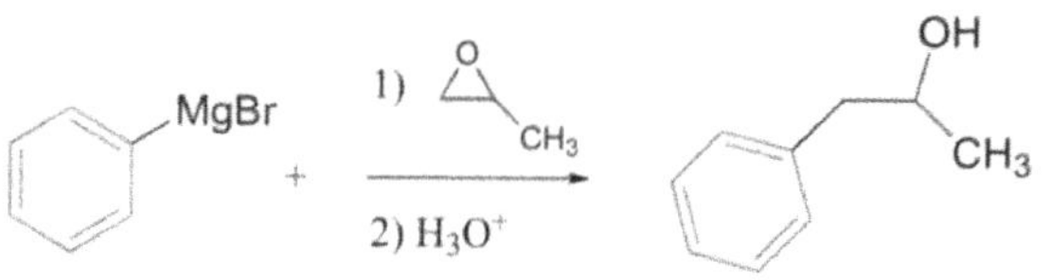

The Grignard reagent (i.e., carbanion in basic conditions) proceeds with the nucleophile (Grignard) attaching the less substituted carbon of the epoxide.

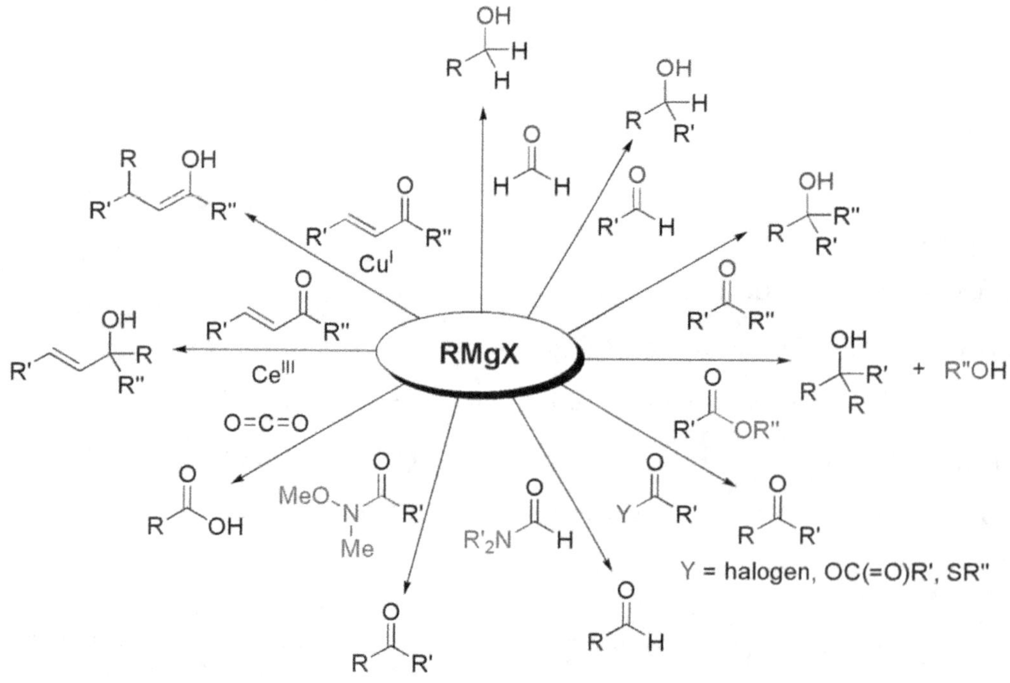

Grignard reaction (RMgX) involving several electrophiles and the resulting products

67. B is correct.

When carboxylic acids and alcohols are combined in the presence of acid catalysts (e.g., H_2SO_4), esters form as the product. Water is produced as a byproduct in the reaction, and its removal from the reaction drives the reaction forward because an equilibrium exists.

68. C is correct.

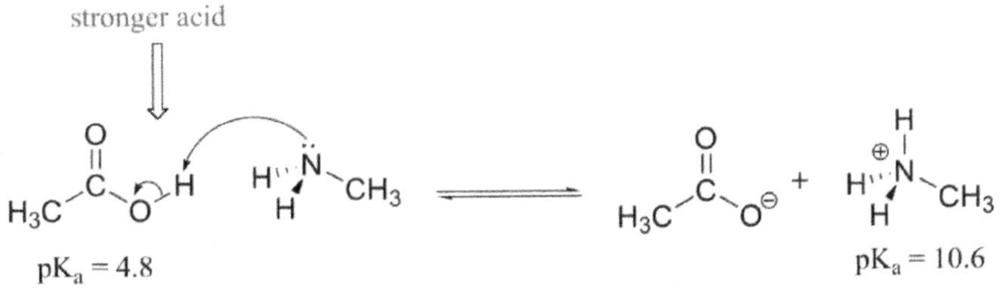

When forming an amide from a carboxylic acid and an amine, higher temperatures are typically needed to drive the reaction forward.

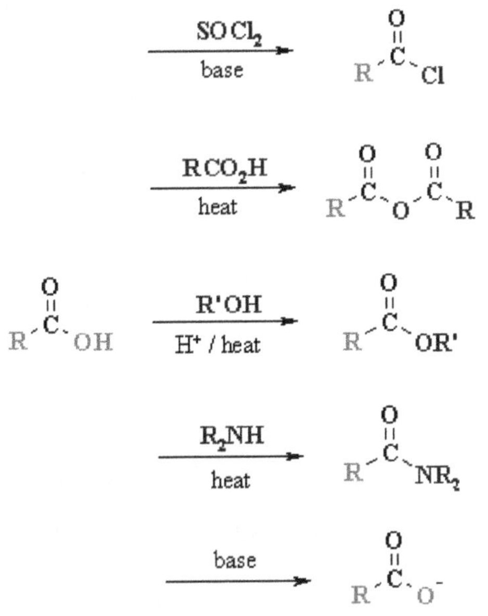

Reactions for a carboxylic acid with specific reagents

The reactivity of the carboxylic acid derivatives is:

acyl chloride > anhydride > carboxylic acid ≈ ester > amide > carboxylate.

An equivalent of water is lost (i.e., dehydration) in the process, so this reaction is a condensation reaction.

69. A is correct.

Although primary and secondary amines can both accept and donate hydrogen bonds, the hydrogen bonding forces are greater for primary amines because these molecules possess two N–H bonds.

Quaternary ammonium salts cannot form hydrogen bonds because the N has four bonds, and no lone pairs of electrons remain on the nitrogen.

70. A is correct.

Propyl substituents (shown below) contain a three-carbon chain.

If the point of attachment (indicated by the squiggle line) is at the second carbon, it is an *iso*propyl group.

B: *tert-* butyl

C: *sec*-butyl

D: isobutyl

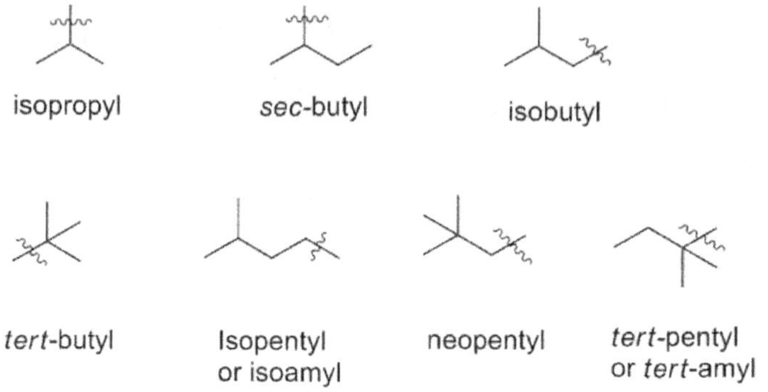

isopropyl

sec-butyl

isobutyl

tert-butyl

Isopentyl
or isoamyl

neopentyl

tert-pentyl
or *tert*-amyl

Common names of alkyl substituents (recognized by IUPAC)

Notes for active learning

Notes for active learning

Diagnostic Test 6 – Answer Key and Detailed Explanations

#	Ans	Topic	#	Ans	Topic
1	D	Nomenclature	36	B	Aromatic compounds
2	B	Covalent bond	37	D	Alcohols
3	A	Stereochemistry	38	A	Aldehydes & ketones
4	A	Molecular structure & spectra	39	C	Carboxylic acids
5	C	Separations & purifications	40	A	COOH derivatives
6	B	Alkanes & alkyl halides	41	D	Amines
7	D	Alkenes	42	B	Nomenclature
8	C	Alkynes	43	B	Covalent bond
9	C	Aromatic compounds	44	D	Stereochemistry
10	D	Alcohols	45	B	Covalent bond
11	D	Aldehydes & ketones	46	D	Stereochemistry
12	A	Carboxylic acids	47	B	Nomenclature
13	B	COOH derivatives	48	B	Covalent bond
14	C	Amines	49	C	Stereochemistry
15	B	Nomenclature	50	A	Molecular structure & spectra
16	D	Covalent bond	51	C	Separations & purifications
17	B	Stereochemistry	52	D	Alkanes & alkyl halides
18	A	Molecular structure & spectra	53	D	Alkenes
19	D	Separations & purifications	54	A	Alkynes
20	D	Alkanes & alkyl halides	55	D	Aromatic compounds
21	C	Alkenes	56	A	Alcohols
22	C	Alkynes	57	B	Aldehydes & ketones
23	A	Aromatic compounds	58	B	Carboxylic acids
24	D	Alcohols	59	C	COOH derivatives
25	B	Aldehydes & ketones	60	D	Amines
26	A	Carboxylic acids	61	C	Nomenclature
27	A	COOH derivatives	62	C	Covalent bond
28	C	Amines	63	C	Stereochemistry
29	D	Nomenclature	64	A	Molecular structure & spectra
30	C	Covalent bond	65	D	Separations & purifications
31	B	Stereochemistry	66	B	Alkanes & alkyl halides
32	B	Molecular structure & spectra	67	D	Alkenes
33	C	Alkanes & alkyl halides	68	A	Alkynes
34	A	Alkenes	69	A	Aromatic compounds
35	B	Alkynes	70	A	Alcohols

This page is intentionally left blank

1. D is correct.

Cyclopropane is the only compound listed that contains three carbons.

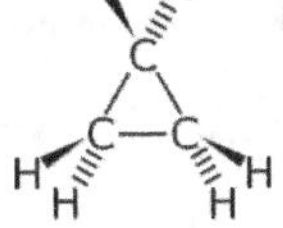

Cyclopropane with the stereochemistry of hydrogens indicated

All other answer choices have four carbons.

2. B is correct. Carbocations are stabilized by resonance and by hyperconjugation.

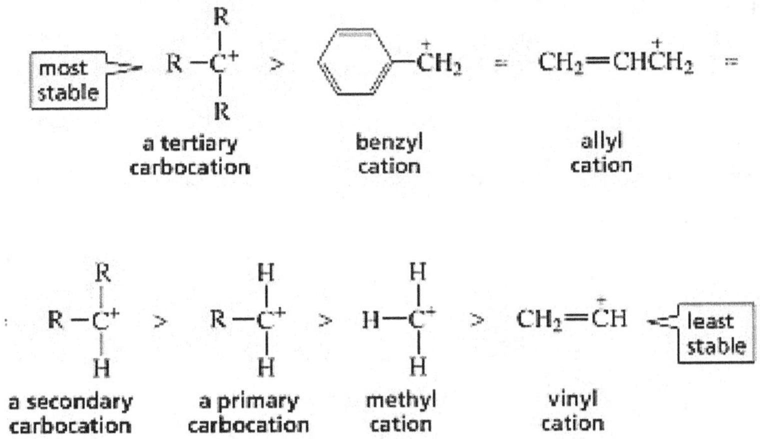

Relative stabilities of carbocations

The phenyl ring offers additional stability due to resonance structures with the delocalization of the *pi* electrons.

The primary non-conjugated carbocation will be the least stable.

3. A is correct.

There are two stereoisomers for the vicinal disubstituted alkene (i.e., *cis* and *trans*).

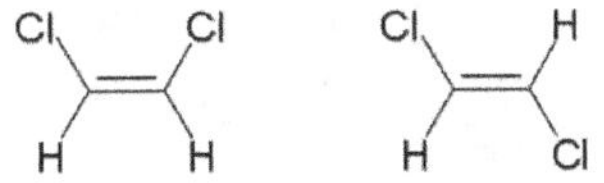

cis- and trans-dichloroethylene

There is one isomer of the geminal (i.e., on the same atom) disubstituted alkene.

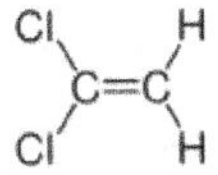

1,1-dichloroethene

4. A is correct.

NMR spectroscopy provides information about the local environment of the proton.

Equivalent hydrogens (i.e., hydrogens in identical locations about other atoms) produce a single NMR signal.

Nonequivalent hydrogens give separate NMR signals on the spectrum.

Since the molecule produces one signal for NMR, all hydrogens are equivalent.

In $(CH_3)_3CCCl_2C(CH_3)_3$, the methyl groups are equivalent, and this molecule produces only one signal in NMR.

B: $(CH_3)_2CHCH_2CH_2CH(CH_3)CH_2CH_3$ produces eight signals.

Additionally, the splitting produces a complex NMR pattern indicating the number of adjacent Hs.

C: $(CH_3)_2CHCH_2(CH_2)_4CH_3$ produces eight signals.

Additionally, the splitting produces a complex NMR pattern indicating the number of adjacent Hs.

D: $CH_3(CH_2)_7CH_3$ produces five signals.

For symmetrical molecules, one signal is for the terminal $CH_3(1,9)$, one for the $CH_2(2,8)$, one for the $CH_2(3,7)$, one for the $CH_2(4,6)$, and one for the CH_2 at 5.

5. C is correct.

For a class of compounds (e.g., hydrocarbons), the lower the molecular weight, the lower the boiling point.

The smallest and most volatile components have the lowest boiling points for a mixture containing compounds of different boiling points.

Molecules with the lowest boiling points (most volatile) vaporize first and travel first up the fractionating column.

Natural gas has the lowest boiling point; therefore, it is isolated first. The next most volatile hydrocarbon is gasoline, and therefore, it has a boiling point that is lower than that of kerosene.

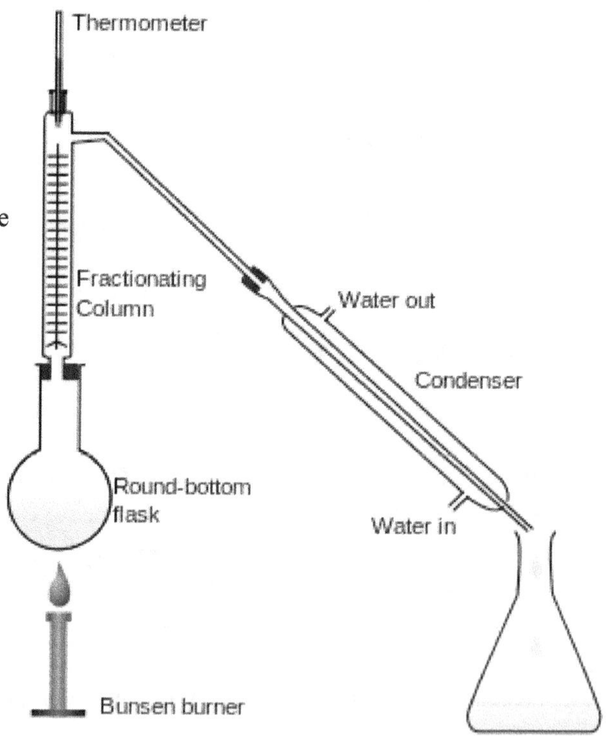

6. B is correct.

Any of six hydrogens (3 on each methyl group) can be substituted to form propyl chloride.

To form isopropyl chloride, the hydrogen extracted must be one of the two on the middle carbon. Based on statistical probability, propyl chloride forms in a 3:1 ratio compared to isopropyl chloride.

A primary radical intermediate is formed when terminal hydrogen is extracted (to form propyl chloride).

When internal hydrogen is extracted (to form isopropyl chloride), a secondary radical intermediate is formed.

Carbon radicals are electron-deficient because they do not have a full octet.

Alkyl groups are electron-donating (via hyperconjugation), so a secondary radical is more stable than a primary radical, and therefore the formation of a secondary radical is more likely.

The reaction proceeding through the secondary radical intermediate is faster. Therefore, the reaction leading to the formation of isopropyl chloride proceeds more readily. This is contrary to strictly statistical considerations, and more isopropyl chloride is formed than the predicted 25%.

Note: it is impossible to predict exactly what percentages form because other factors (e.g., temperature) are important for the empirical yield determination for each product.

7. D is correct.

A *bromonium ion* is a cyclic (three-membered ring) structure of a bromine atom attached to two unsaturated carbons (i.e., an alkene).

The bromonium ion is formed when the nucleophilic double bond adds to a bromine atom of Br_2, releasing Br^-.

The resulting bromine anion (or solvent if it is a nucleophile – contains lone pairs of electrons: H_2O, NH_3 or CH_3OH) bonds to the more substituted carbon of the bromonium structure to give the dibrominated product. Cl_2 follows the same reaction mechanism.

An *anti*-product is formed because the Br^- (or nucleophilic solvent with lone pairs) must approach the three-membered ring of the bromonium ion from the side opposite the bromonium ion.

$CH_3CH_2CH{=}CH_2 + HBr \rightarrow CH_3CH_2CHBrCH_3$: HBr adds to alkenes via a Markovnikov mechanism. First, the double bond attacks the proton, and hydrogen adds to the carbon of the double bond with fewer alkyl substituents (i.e., less substituted carbon).

As a result, a carbocation intermediate forms on the more substituted carbon.

Second, Br^- adds to the carbocation (more substituted carbon) to form the alkyl halide. The mechanism involves a carbocation and not a bromonium ion.

Unlike a bromonium ion, which yields the *anti*-stereochemistry, the trigonal planar carbocation undergoes nucleophilic attack from either face (e.g., the top and bottom). It produces enantiomers (a racemic mixture) if the product contains a chiral center on the carbocation carbon.

8. C is correct.

Using a subscript of n for the number of carbons, the degrees of unsaturation can be determined from the following formulae:

Alkane: C_nH_{2n+2} = 0 degrees of unsaturation

Alkene: C_nH_{2n} = 1 degree of unsaturation

Alkyne: C_nH_{2n-2} = 2 degrees of unsaturation

The formula C_nH_{2n-2} is the general formula for acyclic alkynes.

The general molecular formula for cyclic alkynes is C_nH_{2n-4}.

Cyclic alkynes (i.e., bond angle of 180°) are typically larger-sized rings with at least 8 carbon atoms in the ring.

9. C is correct.

Halides are electron-withdrawing (i.e., deactivating) and are *ortho*/*para*-directing.

Halogens are the exception to the general rule, which states that electron-withdrawing species are deactivating and *meta*-directing.

The halogens (despite being electron-withdrawing due to their high electronegativity) are ortho/para-directing. The lone pair of electrons localize, via resonance, the electron density at the *o/p* positions.

Resonance hybrids show the anion at both *ortho* positions. Note that the resonance structures include an anion at the *para* position.

Therefore, like atoms with lone pairs of electrons attached to the ring, halogens are *ortho-para* directors.

The resonance structure forms a double bond between the halide bearing a formal positive charge and the phenyl ring. Therefore it is not a significant resonance structure and cannot overcome the deactivating (due to electronegativity) effect caused by induction (along the *sigma* bond).

10. D is correct.

For a given class of compounds, the lower the molecular weight, the lower the boiling point.

Alcohols follow the same trend for the boiling point as alkanes.

Ethanol is a two-carbon chain (i.e., lowest molecular weight) with hydrogen bonding (e.g., alcohols and carboxylic acids). It has the lowest molecular weight of all choices and therefore has the lowest boiling point.

11. D is correct.

Ketones and aldehydes have similar chemical properties because both contain a carbonyl group not in conjugation with a heteroatom. They undergo similar reduction and alkylation reactions, although the products of these reactions may be slightly different.

Conversely, aldehydes can undergo an additional oxidation reaction to form carboxylic acids, unlike ketones, which cannot be oxidized further.

12. A is correct.

Acid strength is increased by the inductive effect of electron-withdrawing groups on the neighboring carbon.

Factors that affect the acidity of a molecule:

    1) the more electronegative the substituents, the greater the inductive effect;

    2) the closer the electronegative group is to the carbonyl carbon, the greater the effect;

    3) the greater the number of electronegative substituents, the greater the inductive effect.

Fluoroacetic acid has a $pK_a \approx 2.6$ and is more acidic than others because of the inductive influence of the electronegative fluorine atom.

B: acetic acid (i.e., table vinegar) has a $pK_a \approx 5$ and is the least acidic since it has no electronegative substituents.

C: bromoacetic acid has a $pK_a \approx 3.6$.

D: methoxyacetic acid has a $pK_a \approx 3.6$

13. B is correct.

The reactant benzoyl chloride is an acyl halide that forms a ketone when reacted with an equimolar quantity of a Grignard reagent. Excess Grignard reagent converts the ketone to tertiary alcohol.

    Acyl halide + Grignard reagent → ketone + Grignard reagent → tertiary alcohol

Sample addition reactions of carbonyl compounds and excess Grignard:

    Formaldehyde + excess Grignard → primary alcohol

    Aldehyde + excess Grignard → secondary alcohol

    Ketone + excess Grignard → tertiary alcohol

14. C is correct.

Amine salts are soluble in water because nitrogen has a positive charge and can form ion–dipole interactions with water.

A: nitrogen donates its lone pair of electrons when forming a salt; a positive charge forms on the nitrogen, not a negative charge.

B: the amine salt has a higher molecular weight than the amine, but it is not why the amine salt is more soluble.

D: some amines are soluble in water. Amines follow a similar solubility pattern to carboxylic acids: up to 6 carbons, they are relatively soluble.

15. B is correct.

The longest carbon chain in the molecule is six carbon atoms long.

The two methyl substituents at the second and fourth positions in the chain.

16. D is correct.

The hybridized orbital is a combination of one s and three p orbitals.

Therefore, the energy of the hybridized orbital is between the energy of the combined orbitals.

The s orbital has lower energy than the p orbital.

17. B is correct.

To determine the enantiomer of the compound, draw the mirror image of the compound.

This requires inverting the two stereocenters, as inverting only one results in a diastereomer.

18. A is correct.

The displayed masses on the mass spectrum corresponds to the molecular fragments that have a charge of +1.

It is possible to generate dications from the ionization in the mass spectrometer, and therefore an additional calculation may be required to determine the true mass of the fragment.

19. D is correct.

Fractional (compared to simple) distillation separates molecules with similar boiling points (BP).

Simple distillation is effective for separating molecules with differences in BP of at least 25 °C.

In fractional distillation, a mixture of liquids is heated slowly to separate the more volatile products from less volatile ones (based on differences in boiling points).

The extra surface area in the fractional distillation column forces some less volatile components to condense and reenter the liquid phase, thus enhancing the separation of molecules with differences in boiling points.

20. D is correct.

S_N1 reactions proceed via a carbocation in the first step of the mechanism.

In the second step, the nucleophile forms a new bond by attacking the carbocation.

S_N1 undergoes first-order kinetics, whereby:

$$\text{rate} = k\,[\text{substrate}]$$

where k is a rate constant determined experimentally.

21. C is correct.

cis-3-methyl-2-hexene undergoes Markovnikov addition because the first step is protonation to give the more stable (more substituted) carbonium ion. It gives *syn-* and *anti-*addition products because in the second step, the bromide ion attacks both the top and bottom faces of the carbonium ion.

Adding a hydrogen halide to an asymmetrical alkene leads to either a halogenated alkene in which the halide is on the more substituted carbon or to a product in which the halide is on the less substituted carbon.

The former addition follows Markovnikov's rule.

The latter is an example of an *anti*-Markovnikov addition.

If a mixture of the two products is formed, with a predominance of one product, the reaction is *regioselective*.

22. C is correct.

The higher is the pK_a of a compound, the less acidic the molecule is, and the stronger is the conjugate base.

The pK_a of water is $\approx$ 15-16.

The pK_a of the terminal alkynes is ≈ 25.

The pK_a of the N-H bonds of neutral amines is $\approx$ 36-38.

Therefore, water and the terminal alkyne (but-1-yne or butyne) are more acidic.

23. A is correct.

When 1,3-cyclopentadiene reacts with sodium metal, it is converted from nonaromatic to aromatic.

According to Hückel's rule, a planar cyclic compound is aromatic if conjugated (adjacent sp^2 hybridized carbon atoms) and has 4n + 2 *pi* electrons (where n is any whole number).

24. D is correct.

HBr is a strong acid that protonates the oxygen atom of the alcohol. Then, either the bromide displaces the water to give 1-bromo-2-methylpropane, or a hydride shift occurs to form the tertiary carbocation, which can be trapped by the bromide resulting in 2-bromo-2-methylpropane.

25. B is correct.

Tollens' reagent or Benedict's test are used to detect reducing sugars (i.e., aldehydes).

The aldehyde is oxidized to a carboxylic acid during the reaction.

Monosaccharides are reducing sugars, while some disaccharides, oligosaccharides, and polysaccharides are reducing sugars.

The Tollens' test for aldehydes involves the reduction of silver cations [Ag^+] to reduced silver; the metal precipitates out of solution and coats the inner surface of the reaction flask (i.e., mirror) as a positive test for reducing sugars.

The Tollens' test does not give a positive test with ketones because the ketone is fully oxidized and lacks a C–H bond to be further oxidized and be converted to C–O bonds (e.g., aldehydes to COOH).

26. A is correct.

When dissolved in the basic solution, the carboxylic acid is deprotonated to form the carboxylate salt.

The salt is charged, and this ion can engage in ion–dipole interactions with the solvent and is therefore soluble.

The aldehyde may react with the base to form charged species (either the geminal diol or the enolate), but these charged species are formed reversibly.

Therefore, the acid can dissolve, while the aldehyde is not.

27. A is correct. There are 4 alcohol functional groups present.

A cyclic ester is present. A cyclic ester is a lactone.

$$\begin{array}{c} H \\ | \\ R-C-O-R \\ | \\ OH \end{array}$$

Hemiacetal: alcohol and ether on the same carbon

$$\begin{array}{c} OR'' \\ | \\ R-C-H \\ | \\ OR' \end{array}$$

Acetal: two ethers on the same carbon

28. C is correct.

The most notable intermolecular force for water is hydrogen bonding.

CH_3–CH_2–NH_2 interact with water through hydrogen bonding. Hydrogens, bonded directly to F, O or N, participate in hydrogen bonds. The hydrogen is partial positive (i.e., delta plus or $\partial+$) due to the bond to these electronegative atoms.

The lone pair of electrons on the F, O or N interacts with the $\partial+$ hydrogen to form a hydrogen bond.

CH_3–CH_2–NH_2 is ethylamine. Amines (i.e., nitrogen) can form hydrogen bonds, but thiols (i.e., sulfur) cannot.

29. D is correct.

The longest carbon chain that includes the double bond (i.e., alkene functional group) is a five-carbon molecule.

The chain is numbered with the alkene given the lowest number.

An ethyl (i.e., 2 carbon) substituent is at the second position in the carbon chain.

30. C is correct.

The two allylic cations are resonance forms of each other.

Resonance involves the movement of conjugated *pi* systems, not of the atoms.

A and B are constitutional isomers (i.e., same molecular formula, but different connectivity).

31. B is correct.

All *meso* compounds are molecules with chiral centers and require an inversion center or a plane of symmetry.

These elements of symmetry make molecules with asymmetric carbon atoms achiral overall.

32. B is correct.

The two nuclear spin states for protons are *alpha* and *beta*.

The *alpha* spin state has less energy than the *beta* spin state because the *alpha* spin state has the same direction as the applied external field.

33. C is correct.

Conformations (i.e., rotations around single bond) in which the single bonds are staggered are more stable than those in which they are eclipsed due to steric repulsion and torsional strain (i.e., bonding electron between adjacent atoms – such as observed in eclipsed Newman projections).

Torsional strain is present in Newman projections when the substituents are eclipsed. It originates from the repulsion of the bonding electrons (i.e., not the steric interactions of the substituents).

Conformations that put the largest substituents in an *anti* (180° offset) arrangement are more stable than those in a gauche (60° offset) arrangement.

1,2-dibromoethane is shown anti in the Newman projection

34. A is correct.

Addition reactions require two or more reagents to combine to form a single product.

For the oxidation of an alkene (e.g., epoxidation or dihydroxylation), only the oxygen atoms of a reagent are transferred to the alkene.

A reagent byproduct is typically given off in the reaction, such as carboxylic acid from *m*CPBA oxidation (i.e., peroxy acid) or the reduced metal from potassium permanganate ($KMnO_4$) or osmium tetraoxide (OsO_4) oxidation.

Ozonolysis (O_3 or Cr_2O_7) is a type of oxidation of alkenes that results in splitting the alkene into two separate carbonyl-containing products (i.e., cleavage reaction) and is not an addition reaction.

35. B is correct.

When an alkyne is reduced by sodium in liquid ammonia, a single electron is transferred to the alkyne to produce an anion. The intermediate is a vinyl anion because it is on the atom that is in the double bond.

The anion protonates to give the vinyl radical.

The only cations produced during the reaction are Na^+.

36. B is correct.

Carbonyl compounds are *meta*-directing in EAS reactions.

Since the nitro and carbonyl groups are deactivating, disubstitution is slow, a single *meta* substitution is predominant.

37. D is correct.

The reaction between propanol and PBr_3 (phosphorous tribromide) proceeds via an S_N2 reaction mechanism.

An addition-elimination sequence occurs to the P–Br bond, then substitution by Br^- of the alcohol gives the inverted (*R/S*) product (i.e., bromopropane).

38. A is correct.

The hydrocyanation (i.e., adding $^-C{\equiv}N$) of ketones and aldehydes is an alkylation reaction.

This alkylation reaction is sensitive to the steric environment of the carbonyl electrophile.

Formaldehyde (shown below) contains two hydrogen substituents and is the most reactive.

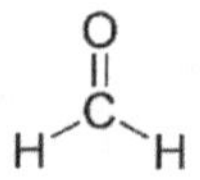

Aldehydes (below) are the second most reactive.

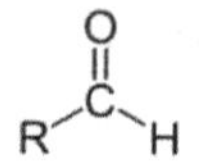

Unhindered ketones (below) are the next most reactive.

Hindered ketones (below) are the least reactive.

These alkyl groups block access to the *pi** orbital of the carbonyl. More sterically-hindered substrates react slowly because the nucleophile (e.g., ⁻C≡N) is impeded in its approach to the delta plus of the carbonyl carbon.

Carbonyl carbon with delta plus carbon (electrophile) indicated

39. C is correct.

Carboxylic acids exchange protons with bases.

When a neutral compound ionizes through the loss of a proton (H^+), the remaining charge of the larger fragment is negatively charged. The proton is positively charged, so the resulting ion is H^+.

The hydroxyl group of a carboxylic acid (or of alcohol) can be replaced (i.e., substitution reaction) with special reagents (e.g., $SOCl_4$ or PBr_3) to form acyl halides (or alkyl halides for alcohol functional groups).

40. A is correct.

In the addition reaction between water and benzoyl chloride, water donates electrons (acting as a nucleophile) to the electrophilic carbonyl of benzoyl chloride.

The chlorine atom, which is not present in the product, acts as a leaving group during the subsequent elimination step whereby the carbonyl carbon reforms.

41. D is correct.

Salts form as solids with tightly compact repeating unit structures, thus requiring more heat to boil than other substances, typically liquids.

42. B is correct.

The longest carbon chain has 5 carbons and is cyclopentane – the root name of the molecule.

It possesses two substituent groups: the chloride and the methyl group.

Stereochemistry is indicated in the structure; the *cis-* notation is necessary for the compound's name.

43. B is correct.

$$H_3C-C(=O)-NH_2$$

acetamide

The resonance structures for carboxylic acids and their derivatives cause the carbonyl to adopt a bond length between a C–O single and a C=O double bond.

Resonance structures of acetamide

The orbitals involved in resonance are the filled, donating *p* orbital of the nitrogen atom and the adjacent electron-accepting carbonyl *pi** orbital.

44. D is correct.

Any molecule that has a non-identical mirror image is chiral and has an enantiomer.

Draw the structure of the cyclic compound:

cis-1,2-dimethylcyclopentane

Comparison of the molecule reveals an internal plane of symmetry.
A *meso* compound and its enantiomer (i.e., a non-superimposable mirror image of a chiral molecule) are the same molecules. Therefore, a *meso* compound does not have an enantiomer.

45. B is correct.

pi bond overlap between the unhybridized orbitals is indicated by the arrow

In organic molecules, the overlap of *pi* bonds is responsible for carbon-carbon *pi* bonds.

The orbitals of double and triple bonds (alkenes and alkynes, respectively) are unhybridized and are more reactive than the *sigma* bonds of single bonds.

Therefore, the reactions of alkenes typically involve breaking the alkene double bond.

46. D is correct.

The number of stereoisomers of a molecule depends on the number of chiral centers, defined as a carbon bonded to 4 different substituents.

The number of stereoisomers is 2^n, where n is the number of chiral centers.

Number the carbon with the carbonyl carbon as #1.

Carbons 1 and 6 are not chiral centers because carbon 1 is attached to only three substituents, while carbon 5 has two identical hydrogen substituents.

Carbons 2, 3, 4, and 5 are chiral centers since each is bonded to 4 different groups.

Therefore, the molecule has 4 chiral centers, and the number of stereoisomers is $2^4 = 2 \times 2 \times 2 \times 2 = 16$.

47. B is correct.

The longest carbon chain in the molecule has 4 carbon atoms, and therefore the root is *but–*.

An alkene is positioned at the second carbon with a suffix of *–ene*.

48. B is correct.

The allylic cation is a carbocation that is one *sigma* bond away from a double bond (i.e., alkene).

Resonance hybrids of an allylic carbocation

Methylene groups are points of saturation in the molecule that can prevent the conjugation of nearby alkenes.

is more stable than

is more stable than

49. C is correct.

The naming of the compounds reveals that the compounds are the same:

7-ethyl-4-isopropyl-3,6-dimethyldecane

50. A is correct.

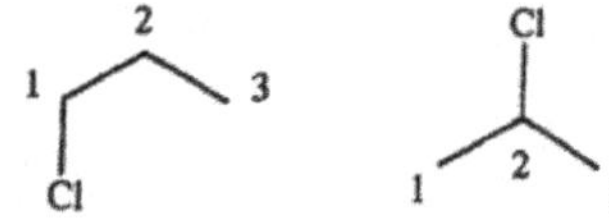

51. C is correct.

Extraction is an organic separation technique that relies on the relative acidity of the molecules.

Aniline is the most basic of the three because it has an amino group, whereby the nitrogen has a lone pair of electrons available for donation to a Lewis acid.

Para-nitroaniline is less basic because the nitro group is electron-withdrawing. The amino group with decreased electron density is less able to donate electrons as needed for a Lewis base.

Nitromethane is not basic because the nitro group is electron-withdrawing.

Acidic extraction makes organic bases soluble in the aqueous phase via protonation of the base to form a soluble ionic salt. Extraction with a sufficiently weak aqueous acid protonates only the most basic compound (aniline), which allows it to migrate into the aqueous phase.

A subsequent extraction with a stronger acid then protonates the less basic (*p*-nitroaniline) of the remaining two molecules, entering the aqueous phase. The third molecule (nitromethane) remains alone in the organic layer, and distillation completes the separation.

If a strong acid were used initially, both aniline and nitroaniline would be protonated and transfer to the aqueous layer, so the extraction (separation) would not be effective.

Extraction with a base is ineffective because aniline and nitroaniline are bases and inert to treatment by a base.

52. D is correct.

Only the electrophile concentration controls the reaction rate because the rate-determining step is the unimolecular formation of the carbocation.

S_N1 reactions proceed via a carbocation in the first step of the mechanism.

In the second step, the nucleophile forms a new bond by attacking the carbocation.

S_N1 undergoes first-order kinetics, whereby:

$$\text{rate} = k[\text{substrate}]$$

53. D is correct.

Of the given molecules, HI is the strongest acid (i.e., forms the most stable anion), so it protonates the fastest (followed by HBr, then HCl).

Also, I^- is the best nucleophile, so it adds to the carbocation most rapidly.

54. A is correct.

One *sigma* bond and two *pi* bonds are used to make the triple bond of alkynes.

Nitriles are another functional group that contains a triple bond.

55. D is correct.

Because the nitrogen lone pair of the aniline is protonated, the group cannot donate electron density through the *pi* system of the aromatic ring.

Furthermore, the nitrogen atom has a formal charge of +1, and this group can act as a deactivating group because of its inductive effect.

56. A is correct.

No carbon-oxygen *pi* bonds exist in alcohols (hydroxyl) functional groups.

Carbonyl groups (i.e., carbon double bonded to oxygen) indicate ketones, aldehydes, or carboxylic acid derivatives (acyl halide, anhydride, ester, and amide).

57. B is correct.

For a ketone converted to an enolate, a base strong enough to remove a proton from the alpha position is used.

If hydroxides or alkoxides are used instead of lithium amide bases, an equilibrium exists between both conjugate forms of the ketone.

Methyllithium and diethylamine may add to the ketone instead of deprotonating it; they are too weak to deprotonate the ketone.

58. B is correct.

Carboxylic acids are molecules that contain a terminal carbon atom with three bonds to oxygen.

It consists of a hydroxyl ($\sim$OH) group and a carbonyl ($\sim$C=O) group.

59. C is correct.

The hydrolysis of an ester is essentially the reverse of Fischer esterification. To drive the reaction forward, excess water and an acid catalyst are used.

The ester hydrolyzes to its two simpler components: the corresponding alcohol and carboxylic acid.

The *R* components of the alcohol product and the alkoxy group of the ester are the same.

60. D is correct.

Arenediazonium salts synthesized from primary aromatic amines are compounds with an N_2^+ group attached to the aromatic ring. They are useful for synthesizing many compounds because they can easily undergo replacement reactions, releasing molecular nitrogen. A nucleophilic substituent attaches to the aromatic ring in its place.

For example, in the presence of cuprous halides, diazonium salts release molecular nitrogen and form halogen-substituted arenes – the *Sandmeyer* reaction. Similarly, nitrogen is released and replaced by the cyanide ion in cuprous cyanide, thus forming aromatic nitriles. In the same way, in cold aqueous solutions, the hydroxyl group replaces the nitrogen to form phenol.

Therefore, compounds I, II, and III can be obtained by replacing nitrogen in arenediazonium salts.

61. C is correct.

The longest carbon chain is composed of five carbon atoms.

The highest priority group is an aldehyde.

The substituent groups are the hydroxymethyl group, the ethyl group, the alkene, and the alkyne.

62. C is correct.

Propene is an example of an alkene, which are molecules with bond angles of about 120 degrees. This molecular geometry affords the substituent groups the greatest amount of spatial separation to minimize the intramolecular Van der Waals repulsions among them.

63. C is correct.

The bromine substitution changes from vicinal (on adjacent carbons) to geminal (on the same carbon). Because the connectivity changes, the structures are constitutional isomers.

The first molecules cannot be chiral because it has a mirror plane of symmetry.

64. A is correct.

The n to *pi** transition for ketones is the electron transition requiring the least energy.

Antibonding orbitals are vacant and serve as the acceptor orbitals for the electron excitation. The orbitals of nonbonding electrons have more energy than the orbitals for *pi* bonds.

Electron transitions from the *sigma* bond are difficult because the electron energy is low and requires high energy to be promoted.

65. D is correct.

Under basic conditions, the caffeine molecule is deprotonated and neutral. As a result, it dissolves in organic solvents (CH_2Cl_2) rather than in H_2O.

Under basic conditions, heptanoic acid is deprotonated, negatively charged (i.e., an anion), and soluble in H_2O.

Continued below

Therefore, in a basic solution, caffeine (i.e., neutral) is miscible in the organic layer, while heptanoic acid (i.e., negatively charged anion) is miscible in the aqueous layer.

Under acidic conditions, caffeine is charged because the NH_2 amine group is protonated, while the heptanoic acid is neutral.

Therefore, the positively charged caffeine is more soluble in H_2O (polar) than in CH_2Cl_2.

Under acidic conditions, heptanoic acid is more soluble in CH_2Cl_2 (nonpolar dichloromethane) when the acid is protonated because heptanoic acid is neutral.

66. B is correct.

Alkanes are only composed of hydrogen and carbon atoms.

Because the electronegativity of these atoms is quite similar, large bond dipoles are not expected.

Hydrogen bonding is a force that acts through highly polarized bonds to form intermolecular bonds to partial positive hydrogen atoms (H attached to F, O or N).

67. D is correct.

In 1,3-butadiene, C_2 and C_3 are sp^2 hybridized with unhybridized p orbitals involved in π bonding.

The bond between C_2 and C_3 is a result of the overlap of two sp^2 hybridized orbitals.

There cannot be a partial double-bond character due to σ electrons; double bonds result only from π electrons.

68. A is correct.

Using a subscript of n for the number of carbons, the degrees of unsaturation can be determined from the following formulae:

Alkane: C_nH_{2n+2} = 0 degrees of unsaturation

Alkene: C_nH_{2n} = 1 degree of unsaturation (1 ring or 1 double bond)

Alkyne: C_nH_{2n-2} = 2 degrees of unsaturation (2 double bonds, 1 double bond and 1 ring or 2 rings)

69. A is correct.

Methyl substituents on benzene (toluene) is an *ortho-* / *para*-director and an activator (rate of the electrophilic aromatic substitution is faster than for benzene)

Friedel-Crafts alkylation reactions work best with electron-rich arenes.

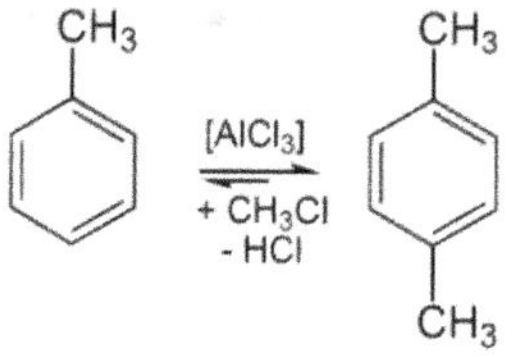

Friedel-Crafts alkylation of toluene

70. A is correct.

It is helpful to draw the structure of known compounds or at least the product of the reaction.

Ethanoate is an ester that has an *n*-propoxy substituent.

Therefore, the alcohol that should be used is 1-propanol.

The remaining component is ethanoic acid.

Notes for active learning

Notes for active learning

Diagnostic Test 7 – Answer Keys and Detailed Explanations

#	Ans	Topic	#	Ans	Topic
1	A	Nomenclature	36	D	Aromatic compounds
2	C	Covalent bond	37	B	Alcohols
3	D	Stereochemistry	38	D	Aldehydes & ketones
4	B	Molecular structure & spectra	39	D	Carboxylic acids
5	D	Separations & purifications	40	C	COOH derivatives
6	D	Alkanes & alkyl halides	41	D	Amines
7	D	Alkenes	42	A	Nomenclature
8	B	Alkynes	43	D	Covalent bond
9	A	Aromatic compounds	44	C	Stereochemistry
10	B	Alcohols	45	C	Aldehydes & ketones
11	D	Aldehydes & ketones	46	A	Carboxylic acids
12	B	Carboxylic acids	47	A	COOH derivatives
13	D	COOH derivatives	48	B	Amines
14	A	Amines	49	B	Nomenclature
15	A	Nomenclature	50	D	Covalent bond
16	C	Covalent bond	51	C	Stereochemistry
17	B	Stereochemistry	52	B	Molecular structure & spectra
18	B	Molecular structure & spectra	53	A	Alkanes & alkyl halides
19	C	Separations & purifications	54	A	Alkenes
20	D	Alkanes & alkyl halides	55	D	Alkynes
21	A	Alkenes	56	D	Aromatic compounds
22	D	Alkynes	57	C	Alcohols
23	A	Aromatic compounds	58	D	Aldehydes & ketones
24	C	Alcohols	59	C	Carboxylic acids
25	A	Aldehydes & ketones	60	D	COOH derivatives
26	A	Carboxylic acids	61	D	Amines
27	A	COOH derivatives	62	D	Nomenclature
28	B	Amines	63	D	Covalent bond
29	B	Nomenclature	64	B	Stereochemistry
30	C	Covalent bond	65	C	Carboxylic acids
31	D	Stereochemistry	66	A	COOH derivatives
32	C	Molecular structure & spectra	67	A	Amines
33	B	Alkanes & alkyl halides	68	A	Nomenclature
34	B	Alkenes	69	D	Covalent bond
35	B	Alkynes	70	A	Stereochemistry

This page is intentionally left blank

1. A is correct.

The molecule has four carbon atoms and two alkenes, hence the root name butadiene.

The double bonds of alkenes are in the first and third positions of the carbon chain.

2. C is correct.

Multiple bonds and rings introduce degrees of unsaturation.

Using a *subscript* of n for the number of carbons, the degrees of unsaturation can be determined from the following formulae:

Alkane: C_nH_{2n+2} = 0 degrees of unsaturation

Alkene: C_nH_{2n} = 1 degree of unsaturation

Alkyne: C_nH_{2n-2} = 2 degrees of unsaturation

Rings = 1 degree of unsaturation

Double bonds = 1 degree of unsaturation

There is *one degree of unsaturation* for this compound (the ring).

Therefore, with no other degrees of unsaturation present, there are no alkenes or double bonds in the molecule.

3. D is correct.

The number of each kind of atom must be consistent among a pair or group of isomers, meaning that the oxidation state of the carbon atoms of the compound is consistent.

A change in the oxidation state of the compound suggests that the hydrogen count is different.

For instance, the isomer of the given alcohol is another alcohol and not an aldehyde or a ketone.

4. B is correct.

The various peaks between 900 and 1500 cm–1 correspond to the unique fingerprint region in the IR spectrum.

The prominent peak at 1710 cm^{-1} indicates the carbon-oxygen double bond of a carbonyl group (aldehyde, ketone, acyl halide, anhydride, carboxylic acid, ester, or amide).

A similar (prominent and broad peak) between 3300 and 3500 cm^{-1} is characteristic of alcohol.

The alcohol of a carboxylic acid would show an IR absorption peak around 2800 and 3200 cm^{-1}.

5. D is correct.

Higher temperature increases the solubility of most substances because the increase in the kinetic energy of higher temperatures allows the solvent molecules to break apart the solute molecules held by intermolecular attractions (i.e., hydrogen bonds, dipole-dipole, and hydrophobic interactions).

For example, a compound is more soluble at 40 °C than at 10 °C. The technique of recrystallization relies upon the principle that a molecule is more soluble at a higher temperature. Therefore, solubility increases with increasing temperature.

The statement "like dissolves like" refers to compounds that are either polar (or ionic) and dissolve in solvents that are polar (or ionic) or non-polar (hydrophobic) and dissolve in non-polar solvents. Therefore, molecules and solvents with similar polarities (ionic or hydrophobic) have increased solubility.

The density of the solvent is independent of its ability to dissolve a given solute.

6. D is correct.

Structural isomers have the same molecular formula (C_7H_{16}) but different atomic connections. Molecular mass would be a consideration if the molecules were not isomers (same molecular formula).

Branching in alkanes lowers the boiling point because branched molecules cannot interact as effectively as unbranched molecules and have less surface area.

2,2,4-trimethylpentane has the lowest boiling point because it is the most highly branched.

Hydrogen bonding is a strong, attractive force between molecules (intermolecular force) but requires hydrogen to be attached to an electronegative atom (fluorine, oxygen, or nitrogen).

The next most attractive intermolecular force is *dipole-dipole interactions*.

7. D is correct.

Geometric isomers have the same molecular formula but different connectivity between the atoms due to the orientation of substituents around a carbon-carbon double bond (or ring).

In *cis* isomers, the same substituents are on one side of the double bond or ring, while in *trans* isomers, the identical substituents are on opposite sides of the double bond or ring.

Isomers (same molecular formula, but different molecules) with no double bonds cannot be geometric isomers.

Isomers include:

> *Constitutional isomers* – different connectivity of the backbone or containing different functional groups.

> *Enantiomers* – mirror images of chiral molecules.

> *Diastereomers* – non-mirror chiral molecules with 2 or more chiral centers or geometric isomers containing double bonds (i.e., designated as *cis/trans* or *E/Z*).

8. B is correct.

Although hydroxide and hot temperatures are employed in this reaction, the potassium hydroxide base is not strong enough to catalyze the isomerization of the triple bond, which is why the internal, not terminal, alkyne is recovered from the reaction.

9. A is correct.

The two arenes activated by nitrogen heteroatoms are more nucleophilic.

The amide does not donate as strongly as the amine, so ring 1 is less nucleophilic than ring 3.

Ring 2 is only activated by the alkyl group and is the least reactive of the three.

10. B is correct.

Ethers follow the same trend as alkanes, so dihexyl ether has the highest boiling point because it has the greatest molecular weight and has the largest surface area.

11. D is correct.

Benedict's test (or Tollens' reagent) detects reducing sugars.

The aldehyde is oxidized to a carboxylic acid in the process.

Monosaccharides are reducing sugars, while some disaccharides, oligosaccharides, and polysaccharides are reducing sugars.

Reducing sugars (and alpha hydroxyl ketones) give a positive Benedict's test: a brown precipitate forms (as does for Fehling's solution).

Tollens' reagent forms silver ions (mirror) as a positive test for reducing sugars.

The Tollens' test for aldehydes involves the reduction of silver cations [Ag^+] to reduced silver; the metal precipitates out of solution and coats the inner surface of the reaction flask.

The ethyl formate has a terminal (HC=O) group, but this group does not oxidize in the Tollens' test conditions because the group is an ester and not an aldehyde. The C–H of the formate ester cannot be oxidized to an O–H bond.

The mechanism for the oxidation of carbonyl C–H bonds typically proceeds through either the hydrate or nucleophilic attack on the carbonyl carbon atom, both of which are not easily done because of the presence of the ester alkoxyl group.

The ester is much less electrophilic than the aldehyde, so it is less reactive.

12. B is correct.

Molecules that possess the hydroxyl group are alcohols and carboxylic acids.

13. D is correct.

Nitriles (−C≡N) have the prefix *cyano* and like alkynes, both functional groups contain triple bonds. However, the nitrile possesses an *sp* hybridized nitrogen atom, giving rise to its alternate chemical reactivity (pK_a = 25).

Nitriles are more electron deficient than alkynes and are more susceptible to nucleophilic attack by molecules such as organolithium compounds and Grignard reagents.

Alkynes undergo addition reactions with strong electrophiles, such as bromine.

14. A is correct.

Carbonyl groups (~C=O) indicate aldehydes, ketones, carboxylic acid, or the four carboxylic acid derivatives (i.e., acyl halide, anhydride, ester, and amide).

Amines do not contain carbonyl groups.

15. A is correct.

The carboxylic acid (~*oic* acid) is considered the 1 position.

The hydroxyl group is in the *ortho* or the 2 position.

16. C is correct.

Hydrogen atoms only bond with other atoms using their 1s orbital.

The nitrogen atoms hybridize their atomic orbitals to produce *sp³* orbital hybrids for bonding.

sp³ orbitals are used because the nitrogen atom forms bonds of equal length with 4 other atoms.

17. B is correct.

There are two stereoisomers for the vicinal disubstituted alkene (1,1-chlorofluoroethene), and there is only one isomer of the 1,2 disubstituted alkene (*cis* and *trans*).

 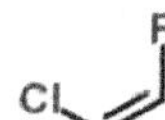

18. B is correct.

Spectroscopy (e.g., NMR) generally is used in the identification of compounds.

Distillation, crystallization, and extraction are commonly used techniques to isolate and purify compounds.

19. C is correct.

Fractional distillation would be the only reasonable method because each product has a different boiling point.

The products would have a similar polarity, so it is difficult to separate by chromatography or extraction.

20. D is correct.

S_N2 reactions are concerted reactions that have a single step and do not involve charged intermediates.

Carbocation intermediates are a feature of S_N1 and E_1 mechanisms.

21. A is correct.

The methyl group adjacent to the carbocation migrates via a methide shift to give a tertiary carbocation, which then loses a proton to form the most substituted alkene (i.e., both carbons of the alkene are tertiary).

22. D is correct.

The oxymercuration-demercuration $[Hg(OAc)_2]$ of alkynes occurs with *Markovnikov addition* to generate a ketone enol.

This enol tautomerizes to form a ketone.

23. A is correct.

The reaction is a Friedel-Crafts acylation using acetic anhydride $(CH_3CO)_2O$ as an acylating agent. The product is methyl ketone.

All halogens are deactivators but direct *ortho / para* because the halogen (like *ortho- / para-* directors) has lone pairs of electrons (on the atom attached to the ring).

The lone pair helps stabilize the positive charge on the ring present in the resonance hybrid intermediates.

In general, deactivators (except for halogens) are *meta*-directors.

24. C is correct.

The notable intermolecular force for water is hydrogen bonding. Alcohols donate and receive hydrogen bonds.

Thiols (S–H bonds) have much smaller dipoles than O–H bonds, so they do not participate in hydrogen bonding.

25. A is correct.

The *first step* in the reaction is coordinating the carbonyl oxygen atom with the positively charged counter ion of the reducing reagent and introducing the hydride to the group.

The *second step* involves the protonation of the alkoxide (negative oxygen atom) intermediate with a weak acid.

26. A is correct.

In addition to the presence of polar functional groups, the molecular weight of a compound is an important influence on its boiling point.

Because carboxylic acids are compared, the one with the highest molecular weight has the highest boiling point.

27. A is correct.

Esters are similar in structure to carboxylic acids, but instead of having a hydroxyl group bonded to the carbonyl carbon atom, an alcohol-derived group is present.

These groups are less electrophilic than ketones and aldehydes, and they have the same oxidation state as carboxylic acids.

Important esters in the body include triacylglycerides or triglycerides.

28. B is correct.

A quaternary ammonium salt has no lone pairs of electrons on the nitrogen atom and cannot function as a nucleophile.

1°, 2° or 3° amines can attack alkyl halides by donating their lone pairs of valence electrons.

Ammonium salts have no available valence electrons and can act as acids (proton donors) but not nucleophiles.

29. B is correct.

The word *acetone* has two important components. The *ace~* is like *acetyl* and indicates that the structure includes a methyl group ($\sim CH_3$) bonded to a carbonyl ($\sim C{=}O$) group.

The suffix *~one* indicates that the carbonyl group is a ketone.

30. C is correct.

The *bond dipole* in H–F is the largest because of the large difference in electronegativity between hydrogen and fluorine. The electrostatic attraction pulls the atoms closer, so the bond is the shortest and the strongest.

The bond dipole in H–I is the smallest because the electronegativity between the atoms is the lowest; therefore, H–I bond is the longest and weakest.

31. D is correct.

The rotation rule (clockwise = *R*, counterclockwise = *S*) determines the configuration after assigning priorities.

When the lowest priority group is on the horizontal bond, the assigned configuration is reversed.

32. C is correct.

For carboxylic acid ($\sim COO\sim$) derivatives, the presence of the heteroatom either increases or decreases the stretching frequency of the carbonyl group.

If the lone pair conjugation with the carbonyl is the more dominant effect (e.g., amides), the carbonyl stretching frequency is lower than 1710 cm^{-1}.

If the *inductive effect* of the heteroatom outweighs the conjugation effect into the carbonyl group (e.g., esters and acid chlorides), then the stretching frequency is higher than 1710 cm^{-1}.

33. B is correct.

The *twist-boat* is at a local energy minimum (or trough) for the conformers of cyclohexane.

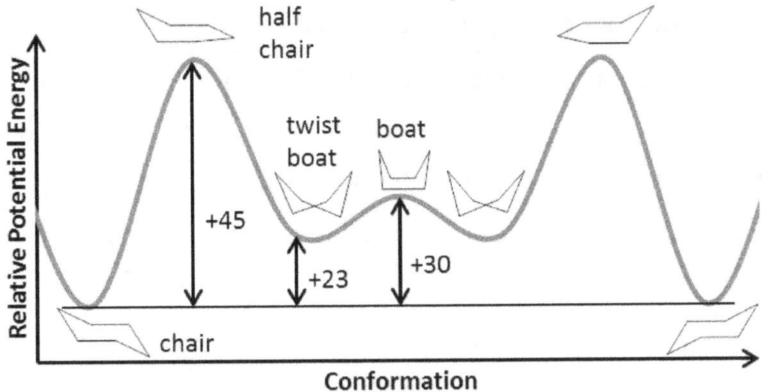

Relative energy diagram for conformers (i.e., chair flips) of cyclohexane

34. B is correct.

The reaction conditions are for halohydration (i.e., halohydrin formation) of an alkene.

The bromonium ion forms first; this is attacked (ring opens) by water as a nucleophile on the more substituted side.

35. B is correct.

Hydrogenation (i.e., reduction) involves the addition of hydrogens (H_2) to an unsaturated molecule.

Catalytic hydrogenation (H_2 / Pd or Pt) of an alkyne is susceptible to a further reduction to yield an alkane.

Alkynes can be reduced to stereospecific products with special reagents.

An alkyne yields a *cis* alkene which requires the Lindlar catalyst (i.e., H_2, Pd, $CaCo_3$, quinolone, and hexane) and a *trans* alkene with Ni (or Li) metal over NH_3 (*l*).

A: *oxidation* is an increase in the number of bonds to oxygen.

Increasing the number of bonds to oxygen often results from decreasing the number of bonds to hydrogens.

C: adding H_2 (i.e., hydrogenation) is an addition, not a substitution, reaction.

D: a *hydration* reaction adds water to an unsaturated (i.e., alkene or alkyne) molecule.

36. D is correct.

In *electrophilic aromatic substitution* (EAS), the aromatic ring acts as a nucleophile attacking an electrophile.

Due to its aromaticity, however, the aromatic ring is relatively unreactive and is a poor nucleophile.

Extremely reactive electrophiles (and the addition of a Lewis base) are used to overcome this limitation.

37. B is correct.

The tosyl group is added with the retention of stereochemistry because the C–O bond is not broken.

Tosylation of secondary alcohol makes it a better leaving group.

The chloride anion is a good nucleophile, displacing the tosylate leaving group to form the secondary alkyl chloride.

D: shows a product that has retained stereospecificity (no inversion).

This retention of stereochemistry occurs when an S_N2 reaction (i.e., inversion) is followed by a second S_N2 reaction.

38. D is correct.

A Michael acceptor is an α, β unsaturated carbonyl – the double bond is between the first and second carbons from the carbonyl carbon.

The best Michael acceptor is the electrophile that best stabilizes the resulting negative charge at the alpha position when a nucleophilic attack (i.e., by Michael donor) occurs at the beta position.

Michael acceptor

Michael donor

Michael adduct

The mechanism for the *Michael reaction* (below):

The Michael donor participates in a nucleophilic attack on the Michael acceptor

39. D is correct.

Carboxylic acids are more acidic than phenols.

The pK_a of carboxylic acids is ≈ 5, and the pK_a of phenols is ≈ 10.

The aryl (on the C=O) proton of benzaldehyde has a pK_a near the lower 40s, while the ketones (e.g., acetone) have pK_a values of ≈ 20.

40. C is correct.

Saponification (below) is the name for the base-promoted hydrolysis of esters.

This type of hydrolysis is typically applied to the formation of soap compounds.

For example, triacylglycerol is a lipid that can undergo saponification, breaking fatty esters with bases, such as sodium hydroxide or potassium hydroxide.

carboxylate ester   sodium hydroxide   sodium carboxylate   alcohol

Base hydrolysis (saponification) of an ester to form a carboxylate salt and an alcohol

41. D is correct.

Ammonia (NH_3) is a basic compound with a low boiling point and exists as a gas at room temperature.

It is basic, colorless, has a pungent smell, and can be toxic if ingested.

42. A is correct.

The names below each benzene derivative are the common names for the benzene-derived compounds (e.g., toluene is the common name for a benzene ring with a methyl substituent).

phenol   toluene   aniline   anisole

nitrobenzene   benzaldehyde   benzoic acid   styrene

benzenesulfonic   acetophenone   *tert*-butylbenzene
acid

Common names for benzene derivatives

For the IUPAC name, the longest carbon chain in the compound is the six-membered benzene ring.

There are two substituent groups present in the molecule: the ethyl group and the methyl group.

Therefore, the IUPAC name is 1-ethyl-3-methylbenzene.

43. D is correct.

Tertiary (3°) carbocations are more stable than secondary (2°) carbocations, which are more stable than primary (1°) carbocations.

Carbocation stability: 3° > 2° > 1° > methyl

The more substituted the cation, the more it benefits from hyperconjugation stabilizing factors.

44. C is correct.

Enantiomers are chiral molecules (i.e., attached to four distinct groups) and are non-superimposable mirror images (i.e., *R* and *S*).

The two molecules are mirror images of each other, so they are enantiomers.

45. C is correct.

Ketones have alkyl or aryl (i.e., ring) substituents on either side of a carbonyl (C=O) functional group.

The IUPAC name for the molecule is butan-2-one or 2-butanone.

46. A is correct.

The carboxylic acid and alcohol starting components for the synthesis of an ester are determined by cleaving the *sigma* bond between the oxygen atom of the alkoxyl (O of the ether) group and the carbonyl (C=O) group.

47. A is correct.

Esters are functional groups that belong to a class of compounds known as carboxylic acid derivatives (along with acyl halides, anhydrides, and amides).

Esters have the same oxidation state as carboxylic acids but have different properties than carboxylic acid groups, such as enhanced electrophilic properties, decreased polarity, and higher pK_a values.

48. B is correct.

Ethylamine is the product since nitrile reduction adds carbon to the alkyl chain.

Cyanide is nucleophilic and can add to methyl bromide, which results in the expulsion of the bromide leaving group. The product of this reaction is cyanomethane.

The nitrile group can be reduced by lithium aluminum hydride ($LiAlH_4$ or LAH) or hydrogenation to form ethylamine.

49. B is correct.

The cyclohexane is the longest carbon chain, and the alcohol is the highest priority group, making the root name "cyclohexanol."

The chlorine and methyl groups are the substituents.

50. D is correct.

The electronegativity difference decides the distribution of electrons between bonded atoms. This unequal distribution of electrons creates the polarity (∂^+ and ∂^-) of the bond.

51. C is correct.

The molecules shown above are mirror images of each other.

Assign *R* and *S* at each chiral center to verify when questions ask the relationship between two chiral molecules.

52. B is correct.

The peak area ratios of the spin states correspond to the values derived from Pascal's triangle.

The heights of the peaks may not correspond to the same ratio, but the area does.

n	2^n	multiplet intensities
0	1	1 Singlet (s)
1	2	1 1 Doublet (d)
2	4	1 2 1 Triplet (t)
3	8	1 3 3 1 Quartet (q)
4	16	1 4 6 4 1 Pentet
5	32	1 5 10 10 5 1 Sextet
6	64	1 6 15 20 15 6 1 Septet
7	128	1 7 21 35 35 21 7 1 Octet
8	256	1 8 28 56 70 56 28 8 1 Nonet

Pascal's triangle

53. A is correct.

The sodium methoxide in this reaction acts as a strong base and abstracts a proton on the carbon atom that neighbors the chloride. The mechanism for this process is E_2.

The reaction is not S_N1 because heat is needed to generate the carbocation, not included in the reaction conditions.

54. A is correct.

In the first step of the reaction, the H^+ adds to the double bond creating a carbocation. The nucleophilic water attacks the carbocation, and the acid is regenerated in the last step of the reaction.

2,3-dimethyl-2-butanol

55. D is correct.

Platinum and palladium are used to hydrogenate alkynes to alkenes (or alkanes) or reduce alkenes to alkanes.

Two moles of hydrogen gas reduces the two pi bonds of the alkyne to an alkane

56. D is correct.

The two arenes (i.e., phenyl moiety) with oxygen substituents are more nucleophilic than methyl.

The ester is less nucleophilic than the aryl ether because of the inductive effect of the oxygen atom on the carbonyl (ester group) that reduces the electron density in the ring.

The lone pair of electrons on the ether resonates into the ring, which increases the electron density for EAS.

57. C is correct.

The 5-carbon molecule of *n*-pentanol has the largest alkyl portion (i.e., the greatest number of London forces). It possesses the hydroxyl group, which allows it to hydrogen bond (increases boiling point).

An overall molecular dipole exists for this molecule, which contributes to its high boiling point.

58. D is correct.

1,3-dithiane serves as a nucleophilic carbonyl equivalent.

Treatment of 1,3-dithiane with n-BuLi generates a carbanion between both sulfur atoms of the molecule, and this anion can add to electrophiles such as alkyl halides.

Both C–H bonds of the 1,3-dithiane can be substituted with alkyl groups.

The hydrolysis of the dithiane to the carbonyl group requires mercury salts in an acidic solution.

59. C is correct.

The initial reaction between the amine and carboxylic acid is neutralization (i.e., proton transfer).

However, at high temperatures, this proton transfer is reversible, and the reaction has enough energy to promote the nucleophilic attack of the carboxylic acid instead, leading to amide formation.

60. D is correct.

Anhydride functional groups (below) have oxygen between two carbonyl carbons.

The anhydride functional group does not exist in the reference molecule.

61. D is correct.

Primary amines have one alkyl group attached to the nitrogen atom; secondary amines have two alkyl groups attached to the nitrogen atom; tertiary amines have three alkyl groups attached to the nitrogen atom.

62. D is correct.

The alcohol of this aromatic compound has the highest priority, and the carbon count starts at this position. The root (and suffix) name of the compound is "phenol."

The ethyl group is numbered as three instead of four because the numbering favors the lower position values.

63. D is correct.

The molecular formula given above is for an acyclic alkyne.

Acyclic alkynes have a linear geometry (i.e., 180 degrees), and the carbon atoms have *sp* hybridization.

Cyclic alkynes have two fewer hydrogen atoms because those bonds are replaced with carbon-carbon bonds to form a ring.

64. B is correct.

In the polarimeter (i.e., plane-polarized light), enantiomers have the same magnitude of specific rotation but the opposite sign.

65. C is correct.

benzoic acid
insoluble in water

sodium benzoate
soluble in water

66. A is correct.

The reaction best suited to occur under "normal conditions," rather than harsh conditions, is the aminolysis of the ester to form an amide.

Amides are less electrophilic than esters, and basic conditions are needed to convert the amide back to the ester.

67. A is correct.

Quaternary ammonium salts can be formed from the *N*-alkylation of tertiary amines.

When tertiary amines are converted to ammonium salts, the inductive effects of the nitrogen atom are larger, and the magnitude of the carbon–nitrogen dipoles increase.

68. A is correct.

The *para-* notation means that two substituent groups are on opposite sides of the aromatic rings in a C_1–C_4 relationship.

When the bromine atoms are adjacent (C_1–C_2), the isomer is *ortho*.

When the bromine atoms are in a C_1–C_3 relationship, the isomer is *meta*.

69. D is correct.

Wohler's experiment is significant because it demonstrated that organic compounds could be created from inorganic compounds.

This result ran counter to the belief that the material that composed life was different from the matter of nonliving things, a theory called *vitalism*.

70. A is correct.

It is important to remember that if an alkene has geminal disubstitution (i.e., two identical moieties bonded to the same alkene carbon), then *cis-* and *trans-*isomerism is not possible.

Notes for active learning

Notes for active learning

Supplemental Questions – Answer Keys and Detailed Explanations

#	Ans	Topic	#	Ans	Topic
1	D	Nomenclature	20	D	Alkanes & alkyl halides
2	D	Covalent bond	21	B	Alkenes
3	C	Stereochemistry	22	C	Alkynes
4	D	Molecular structure & spectra	23	B	Aromatic compounds
5	B	Separations & purifications	24	A	Alcohols
6	C	Alkanes & alkyl halides	25	A	Aldehydes & ketones
7	C	Alkenes	26	D	Carboxylic acids
8	A	Alkynes	27	D	COOH derivatives
9	C	Aromatic compounds	28	C	Amines
10	B	Alcohols	29	A	Nomenclature
11	D	Aldehydes & ketones	30	A	Covalent bond
12	C	Carboxylic acids	31	D	Stereochemistry
13	A	COOH derivatives	32	D	Molecular structure & spectra
14	B	Amines	33	B	Alkanes & alkyl halides
15	C	Nomenclature	34	D	Alkenes
16	B	Covalent bond	35	C	Alkynes
17	D	Stereochemistry	36	B	Aromatic compounds
18	D	Molecular structure & spectra	37	C	Alcohols
19	C	Separations & purifications	38	D	Aldehydes & ketones

This page is intentionally left blank

1. D is correct.

The longest carbon chain is composed of seven carbon atoms.

There are chlorine, ethyl, and methyl substituent in this compound.

2. D is correct.

The benzylic position is one carbon away from benzene or aromatic rings.

Without the aromatic ring, the cation is an allylic carbocation when one carbon is away from a double bond.

Vinyl means on the double bond.

3. C is correct.

Draw the different isomers of butene.

This compound can be drawn with the double bond terminal (the carbons are numbered according to IUPAC).

The terminal carbon in the double bond is numbered 1.

There are two geometric isomers of butene (i.e., *cis*-butene and *trans*-butene).

cis-2-butene   *trans-2-butene*

Geometric isomers are a subset of structural isomers. Geometric requires that the substituents from the double bond can be the same (i.e., *cis* and *trans*) or different (i.e., *E* and *Z*).

1-butene or butene (the position 1 is implied by IUPAC)

Isobutylene (not 2-methyl propene) according to IUPAC

Therefore, there are four structural isomers of butene.

4. D is correct.

The local magnetic field generated by the circulating current of the benzene ring causes the protons attached to the ring to be further deshielded (diagram below).

Electron-donating and withdrawing groups attached to the ring may shift the resonances of specific protons up or downfield.

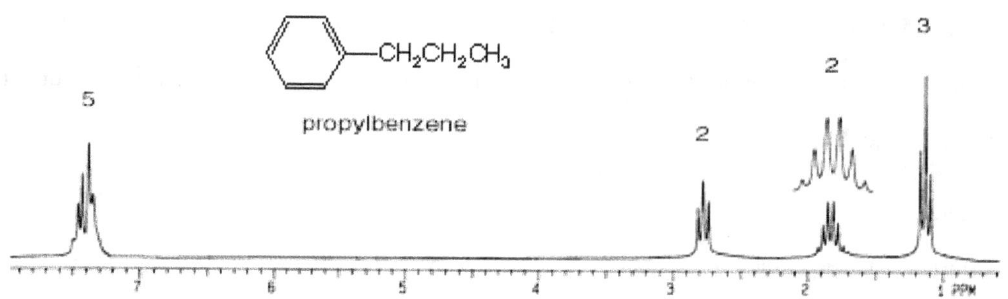

^1H NMR spectra with characteristic absorption for the phenyl ring between 6.0-8.0 ppm

5. B is correct.

Statement I: most substances become more soluble at higher temperatures because solubility constants, like other equilibrium constants (Le Châtelier's principle), are a function of temperature.

Statement II: polar or ionic molecules dissolve in polar solvents, whereas non-polar compounds dissolve in non-polar solvents. Therefore, similar polarities increase solubility.

Statement III: molecular weight of the solvent is not related to solubility.

Statement IV: density of the solvent is not related to solubility.

6. C is correct.

A nucleophile donates lone pairs of electrons; therefore, it is a Lewis base.

7. C is correct.

Although these reactions produce the same products and in the same quantities, what is different is the rate of the borane addition for each alkene.

The *E* has the highest priority substituents on opposite sides, while the *Z* alkene has both alkyl substituents on the same side. Because *Z* alkenes are less thermodynamically stable than *E* alkenes, *Z* alkenes are more reactive.

Furthermore, the approach of the borane to the *Z* alkene is less sterically demanding because the *Z* alkene is more open to an attack (alkyl groups are on the same side).

8. A is correct.

In the presence of excess hydrogen gas, the triple bond of 3-heptyne is completely reduced to the alkane.

Platinum and palladium catalysts can be used to reduce alkynes to alkenes or alkanes.

9. C is correct.

$FeBr_3$ acts as a catalyst to activate the alkyl bromide for electrophilic aromatic substitution (EAS) in this Friedel-Crafts alkylation.

Since the ethyl substituent on ethylbenzene is slightly electron-donating, it functions as an *ortho-*, *para*-director for the substitution reaction.

Due to steric hindrance at the *ortho*-position, *para*-substitution is the major product.

10. B is correct.

Alcohols follow the same trend for boiling points as alkanes, with longer chain molecules having higher boiling points.

Hexanol is the longest chain and, therefore, has the highest boiling point.

11. D is correct.

Ketones are functional groups that Cr^{VI} reagents can no longer oxidize. These groups do not undergo further reactions because there are no C–H bonds to the carbonyl carbon of the ketone.

Aldehydes, however, may have subsequent oxidations to generate carboxylic acids.

12. C is correct.

Carboxylic acids are acidic functional groups that have a hydroxyl group directly bonded to the carbonyl group.

Carboxylic acids are acidic because the conjugate base is stable due to resonance stabilization of the negative charge on oxygen.

The molecule contains three hydroxyl groups and an arene.

13. A is correct.

Amides are functional groups that possess an amino group bonded to a carbonyl moiety.

The reference molecule is a tertiary amine because it is bonded to three R groups.

14. B is correct.

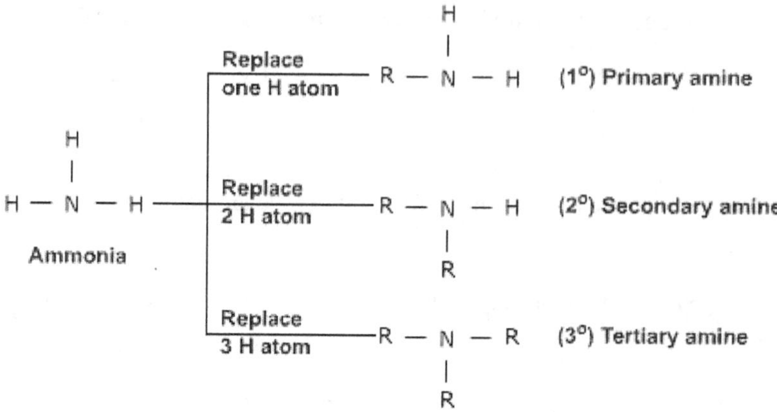

Primary amines are amines that have one alkyl group bonded to the nitrogen atom.

Secondary amines have two alkyl groups bonded to the nitrogen atom.

Tertiary amines contain nitrogen atoms that are bonded to three alkyl groups.

15. C is correct.

The longest carbon chain is composed of six carbon atoms and is the cyclohexene substructure.

The molecule contains a methyl group in the fourth position.

16. B is correct.

The hydrogen atom bonds by overlapping its $1s$ orbital with the orbital of a bonding partner.

The carbon atom is bonded to three atoms and is positively charged, indicating that the carbon atom is sp^2 hybridized.

17. D is correct.

An asymmetric carbon refers to a chiral carbon: carbon bonded to four different substituents.

The other compounds do not contain asymmetric carbons because at least two of the three atoms or groups bonded to each carbon atom are the same.

18. D is correct.

The *shielding effect* arises from the electron density associated with the nuclei.

H nuclei with more electron density are shielded; their chemical shifts appear more upfield in the spectrum.

19. C is correct.

Thin-layer chromatography (TLC) relies upon the principle that a mobile phase (solvent) flows through a stationary phase (sample on the polar plates) and migrates at different rates depending on the attraction towards these phases.

If components within the sample are polar, they adhere to the polar plates and are held immobile, producing a spot near the original location.

If components within the sample are nonpolar, they are weakly attracted to the stationary phase, dissolve in the nonpolar solvent, move up the TLC plate with the solvent, and produce a spot further from the original location.

Each molecule has a reference R_f value that is a fraction of the movement of the sample to the distance of the solvent.

Nonpolar samples move further and have higher R_f values than polar samples, which migrate a shorter distance because they remain bound to the polar plate.

$$R_f = \text{migration distance of a sample/migration distance of the solvent front}$$

20. D is correct.

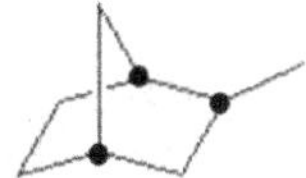

This molecule has only three tertiary alkyl positions (i.e., carbon bonded to three other carbons).

21. B is correct.

The addition of deuterium (an isotope of hydrogen) across a C=C double bond is metal-catalyzed. It proceeds with *syn*-stereoselectivity, adding deuterium to the same face of the double bond.

22. C is correct.

Protonation of alkynes with acids generates the more substituted carbocation because this cation is the more stable intermediate. Because the cation exists on a vinyl carbon atom, this cation is known as a *vinyl cation.*

23. B is correct.

The aromatic stabilization energy of benzene decreases the reactivity of its *pi* system relative to isolated or conjugated/nonaromatic alkenes.

Because of this stabilization, arenes resist metal-catalyzed hydrogenation reactions and may require more specialized catalysts to facilitate their reductions.

24. A is correct.

When phenols are deprotonated (i.e., act as acid), a phenoxide ion (i.e., phenolate ion) is produced, as shown.

The pK_a of phenols is much lower than it is for alcohols.

Methanol has a pK_a of 16, while phenols have pK_a values $\approx$ 9.9.

25. A is correct.

Aldehydes can convert to the enol form, although the equilibrium constant favors the aldehyde.

Conversion to the enol form destroys the stereocenter because this isomer is achiral.

When the enol form converts back to the ketone, protonation may occur on either face of the group, and this interconverts the configuration (i.e., producing R and S isomers).

The process is catalyzed by the presence of an acid or base.

26. D is correct.

A chemical equilibrium exists for this reaction, and water is given off as a byproduct in the process.

27. D is correct.

The *hydrolysis of an ester* is essentially the reverse of Fischer esterification.

To drive the reaction forward, excess water and an acid catalyst are used. The ester hydrolyzes to its two simpler components: the corresponding alcohol and carboxylic acid.

The *R* component of the alcohol product and the alkoxy group of the ester are the same.

28. C is correct.

Amines possess nitrogen atoms with lone pairs that are more basic and nucleophilic than oxygen atoms and halogens such as fluorine.

This is because the nitrogen atom is a less electronegative heteroatom; therefore, the nonbonding valence electrons of nitrogen have more energy and are more reactive.

29. A is correct.

One option is to draw the atoms to determine the atom count.

Alternatively, using a subscript of n for the number of carbons, the degrees of unsaturation can be determined from the following formulae:

> Alkane: C_nH_{2n+2} = 0 degrees of unsaturation
>
> Alkene: C_nH_{2n} = 1 degree of unsaturation
>
> Alkyne: C_nH_{2n-2} = 2 degrees of unsaturation
>
> Rings = 1 degree of unsaturation

The reference molecule has 2 rings and therefore 2 degrees of unsaturation:

> C_nH_{2n-2}
>
> $C_8H_{16-2} = C_8H_{14}$

30. A is correct.

Using the formula for calculating the degrees of unsaturation reveals that 2 unsaturation elements (either rings or *pi* bonds) exist.

Structures that contain atoms with satisfied octets are favored.

Molecules with charged carbon atoms tend to be less stable.

31. D is correct.

The number of stereoisomers for a given molecule depends on the number of asymmetric carbon atoms (or chiral centers) present in the molecule.

There is a 2^N number of possibilities, where N is the number of chiral centers present.

The molecule has 4 chiral centers: $2^4 = 2 \times 2 \times 2 \times 2 = 16$ stereoisomers.

For some molecules, symmetry elements may be redundant structures, so the 2^N calculation gives the maximum number of stereoisomers and not necessarily the actual number that exists.

32. D is correct.

2-methylpropanoic acid (or isobutyric acid)

Carbonyl carbons (~C=O) show an IR absorption between approximately 1630 and 1780 cm^{-1}.

The carbonyl (~C=O) of a carboxylic acid is reported to be between 1710 and 1780 cm^{-1}.

Because carboxylic acids (~COOH) contain a hydroxyl group, another signal around 3300 to 3400 cm^{-1} should be expected in the spectrum for this compound.

33. B is correct.

S_N2 is bimolecular;

rate = k [substrate] × [nucleophile].

Therefore, if the nucleophile ($^-$OH) concentration is doubled, the reaction rate doubles.

Since the alkyl halide is primary, a unimolecular (S_N1) reaction occurs.

Water stabilizes the carbocation intermediate and thereby increases the rate of the reaction.

34. D is correct.

The trapping of mercury-alkene cations typically opens on the more substituted side, but the *tert*-butyl group sufficiently blocks access to this side. Therefore, the cation is opened on the more terminal side.

35. C is correct.

Use the following formulae to calculate the degrees of unsaturation:

C_nH_{2n+2}: for an alkane (0 degrees of unsaturation)

C_nH_{2n}: for an alkene or a ring (1 degree of unsaturation)

C_nH_{2n-2}: for an alkyne, 2 double bonds, 2 rings or 1 ring and 1 double bond (2 degrees of unsaturation).

The general chemical formula for alkynes is C_nH_{2n-2} because each *pi* bond of the alkyne represents one degree of unsaturation.

Cyclic alkynes have an additional degree of unsaturation because of the cyclic structure.

Therefore, C_9H_{16} is the molecular formula that describes an acyclic alkyne.

36. B is correct.

Degenerate orbitals are orbitals that have the same energy.

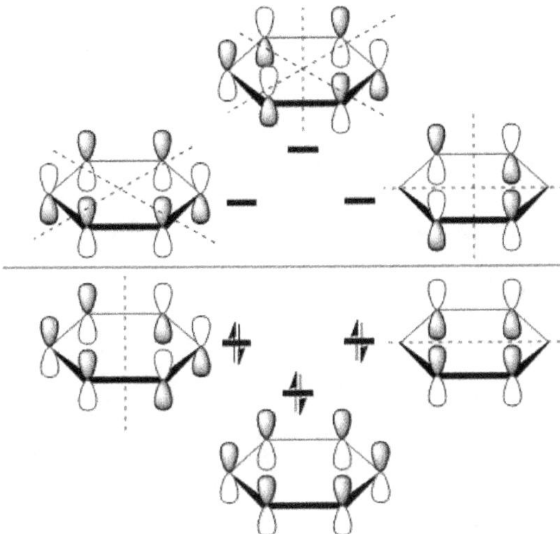

Bonding orbitals are below the horizontal plane, while antibonding orbitals are above

In the molecular orbital (MO) diagram for benzene: MOs $\pi2$ and $\pi3$ are positioned in the second energy level, and $\pi4$ and $\pi5$ MOs are positioned on the third energy level.

Therefore, there are two pairs of degenerate MOs.

37. C is correct.

The formation of an inorganic ester by adding alcohol and phosphoric acid is an example of dehydration synthesis because an equivalent of water is lost during the process.

38. D is correct.

If a stronger reducing agent, such as lithium aluminum hydride ($LiAlH_4$), is used, the aldehyde is reduced to the primary alcohol.

Ester is reduced to an aldehyde with the mild reducing agent of DIBAL

Notes for active learning

Notes for active learning

Notes for active learning

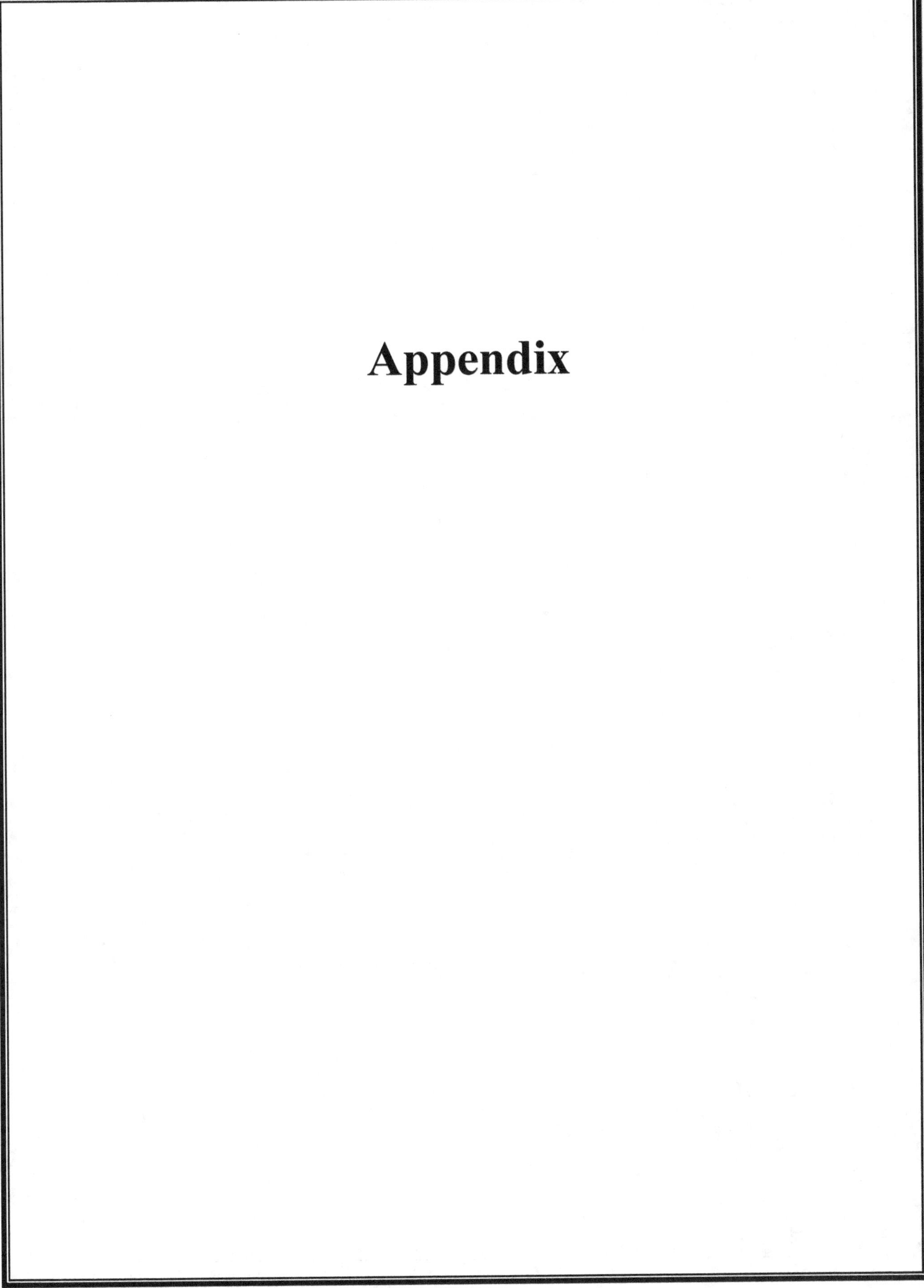

Appendix

Glossary of Organic Chemistry Terms

A

Acetal – product formed by the reaction of an aldehyde with *alcohol*; the general structure of an acetal is:

$$R - \underset{\underset{H}{|}}{C} \begin{smallmatrix} \diagup OR' \\ \diagdown OR' \end{smallmatrix}$$

Achiral (or non-chiral) – the opposite of *chiral*; can be superimposed on its mirror image (e.g., CH_4); do not rotate plane-polarized light.

Acid – an agent able to produce positively charged hydrogen ions (H^+); since the hydrogen ion is a bare proton, it usually exists in a solvated form (such as H_3O^+); a proton donor or an electron pair acceptor; see *Brønsted-Lowry theory of acids and bases*, *Lewis acid* and *Lewis base*.

Acid-base reaction – a neutralization reaction in which the products are salt and water.

Activated complex – molecules at an unstable intermediate stage in a reaction.

Activating group – a substituent that increases the rate of electrophilic aromatic substitution (EAS) when bonded to an aromatic ring.

Activation energy – the minimum energy which reacting species must possess in order to be able to form an "activated complex," or "transition state," before proceeding to the products; the difference in potential energy between the ground state and the transition state of molecules; molecules of reactants must have this amount of energy to proceed to the product state; the activation energy (E_a) may be derived from the temperature dependence of the reaction rate using the *Arrhenius equation*.

Acyl group – a substituent with the following structure, where R can be an alkyl or aryl group:

$$R - \overset{\overset{O}{\|}}{C} -$$

Acyl halide – a compound with the general structural formula:

$$R - \overset{\overset{O}{\|}}{C} - X$$

Acylation – a reaction in which an acyl group is added to a molecule.

Acylium ion – the resonance stabilized cation:

$$R\overset{+}{C} = \overset{..}{O}: \longleftrightarrow RC \equiv \overset{+}{O}:$$

Addition – a reaction that produces a new compound by combining the elements of the original reactants.

Addition elimination mechanism – the two-stage mechanism by which nucleophilic *aromatic* substitution (NAS) occurs; in the first stage, the addition of the nucleophile to the carbon bearing the *leaving group* occurs; an elimination follows in which the leaving group is expelled.

Addition reactions – an unsaturated system is saturated or partly saturated by adding a molecule across the multiple bonds (e.g., adding bromine to ethene to form 1,2-dibromoethane).

Adduct – the product of an addition reaction.

Alcohol – a molecule containing a hydroxyl (~OH) group; also, a functional group.

Aldehyde – a molecule containing a terminal carbonyl (~CHO) group; also, a functional group.

Alicyclic compound – an *aliphatic cyclic* hydrocarbon; a compound contains a ring but not an *aromatic* ring; see *aliphatic compound* and *cyclization*.

Aliphatic compound – a straight-chain or branched-chain hydrocarbon, an *alkane*, *alkene* or *alkyne*.

Alkaloid – organic substances occurring naturally, which are basic, forming salts with acids; the basic group is usually an amino function.

Alkane – a hydrocarbon that contains only single covalent bonds (i.e., containing only C–H and C–C single bonds); the general *alkane* formula is C_nH_{2n+2}.

Alkene – a molecule containing one or more carbon-carbon double bonds; also, a *functional group*.

Alkoxide ion – an anion formed by removing a proton from an *alcohol*; the RO^- ion.

Alkoxy free radical – formed by homolytic cleavage of an *alcohol* ~OH bond; the RO· radical.

Alkyl group – an *alkane* molecule from which a hydrogen atom has been removed; abbreviated as "R" in structural formulas.

Alkyl halide – a hydrocarbon that contains a halogen substituent, such as fluorine, chlorine, bromine, or iodine.

Alkyl-substituted cycloalkane – a cyclic hydrocarbon to which one or more *alkyl group*s are bonded; compare with *cycloalkyl alkane*.

Alkylation – a reaction in which an *alkyl group* is added to a molecule.

Alkyne – a molecule containing one or more carbon-carbon triple bonds; also, a functional group; the general formula is C_nH_{2n-2}.

Allyl group – the $H_2C=CHCH_2$ group contains 3 carbon atoms and a double bond, $C_1=C_2-C_3$, where C_3 is the "allylic position" or "*allylic carbon* atom."

Allylic carbon – a sp^3 carbon adjacent to a double bond.

Allylic carbocation – the $H_2C=CHCH_2^+$ ion.

Allylic rearrangement – the migration of a double bond in a 3-carbon system from carbon atoms one and two to carbon atoms two and three (e.g., $C_1=C_2-C_3-X$ to $X-C_1-C_2=C_3$).

Analogue – in organic chemistry, chemicals that are similar but not identical (e.g., the hydrocarbons are similar, but an *alkane* is different from *alkenes* and *alkynes* because of the types of bonds they contain; therefore, an *alkane* and an *alkene* are analogs).

Angle of rotation (α) – in a polarimeter, the angle right or left in which plane-polarized light is turned after passing through an optically active compound in solution.

Amide – a molecule containing a carbonyl group attached to nitrogen ($\sim CONR_2$); also, a functional group.

Amine – a molecule containing isolated nitrogen (NR_3); also, a functional group.

Anion – a negatively charged ion.

Anomers – the specific term used to describe carbohydrate stereoisomers differing only in configuration at the hemiacetal carbon atom.

***Anti*-addition** – a reaction in which the two groups of a reagent X–Y add on the opposite faces of a carbon-carbon bond.

***Anti*-aromatic** – a highly unstable planar ring system with $4n$ *pi* electrons.

Antibonding molecular orbital – contains more energy than the atomic orbitals (AO) from which it was formed; an electron is less stable in an antibonding orbital than in its original atomic orbital.

***Anti*-conformation** – a type of staggered conformation in which the two big groups are opposite each other in a *Newman projection*.

***Anti*-Markovnikov addition** – a reaction in which the hydrogen atom of a hydrogen halide bonds to the carbon of a double bond that is bonded to *fewer* hydrogen atoms; the addition takes place *via* a free-radical intermediate rather than a carbocation; compare with *Markovnikov rule*.

***Anti*-periplanar (anti-coplanar)** – the conformation in which hydrogen and the leaving group are in the same plane and on opposite sides of a carbon-carbon single bond; also, the conformation required for E_2 elimination.

Aprotic solvents – solvents that do not contain O–H or N–H bonds.

Arene – an *Aromatic* hydrocarbon.

Aromatic – possesses aromaticity; aromaticity is the property of planar (or nearly planar) cyclic, conjugated systems having (4n + 2) conjugated *pi* electrons; the delocalization of the (4n + 2) *pi* electrons give them exceptional stability (aromatic compounds are unusually stable compounds); for benzene, the most common aromatic system (n = 1, therefore 6 *pi* electrons), the aromaticity confers the characteristic reactivity of electrophilic substitution.

Aromatic compound – possesses a closed-shell electron configuration and resonance; obeys *Hückel's rule*.

Aryl – an *aromatic* group as a substituent.

Aryl group – a group, produced by the removal of a proton from an *aromatic* molecule.

Aryl halide – a compound in which a halogen atom is attached to an *aromatic* ring.

Association – a term applied to the combination of molecules of a substance with one another to form more complex systems; see *dissociation* and *dissociation constant*.

Asymmetry – a term applied to an object or molecule that does not possess symmetry.

Asymmetric induction – a term applied to the selective synthesis of one diastereomeric form of a compound resulting from the influence of an existing *chiral* center adjacent to the developing asymmetric carbon atom; this usually arises because, for steric reasons, the incoming atom or group does not have equal access to both sides of the molecule.

Atom – the smallest amount of an element; a nucleus surrounded by electrons.

Atomic mass (A) – the sum of the weights of the protons and neutrons in an atom; a proton and neutron each have a mass of 1 atomic mass unit.

Atomic number (Z) – the number of protons or electrons in an atom.

Atomic 1s orbital – the spherical orbital nearest the nucleus of an atom.

Atomic orbital – a region in space around the nucleus of an atom where the probability of finding an electron is high, which may be described in terms of the four quantum numbers.

Atomic p orbital – an hourglass-shaped orbital, oriented on x, y and z-axes in three-dimensional space.

Atomic s orbital – a spherical orbital.

Avogadro's constant – number of particles (atoms or molecules) in one mole of a pure substance: 6.022×10^{23}.

Axial bond – a bond perpendicular to the equator of the ring (up or down), typically in a chair cyclohexane.

B

Baeyer reagent – cold, dilute potassium permanganate; used to oxidize *alkenes* and *alkynes*.

Base – a substance that can combine with a proton (i.e., a proton acceptor or an electron-pair donor); see *Brønsted-Lowry theory of acids and bases, Lewis acid* and *Lewis base*.

Benzene (Benzenoid) ring – an *Aromatic* ring with a benzene-like structure.

Benzyl group – a *benzene ring* plus a methylene (~CH_2~) unit: $C_6H_5CH_2$.

Benzylic position – the position of carbon attached to a *benzene ring*.

Benzyne – an unstable intermediate consisting of a benzene ring with an additional (triple) bond created by the side-to-side overlap of sp^2 orbitals on adjacent carbons of the ring.

Bicyclic – a molecule with two rings that share at least two carbons.

Bimolecular reaction – two species (e.g., molecules, ions, radicals) react to form new chemical species; most reactions are bimolecular or proceed through a series of bimolecular steps.

Bond angle – the angle formed between two adjacent bonds on the same atom.

Bond-dissociation energy – the amount of energy needed to homolytically fracture a bond.

Bond energy – the energy required to break a particular bond by a homolytic process.

Bond length – the equilibrium distance between the nuclei of two atoms or groups bonded to each other.

Bond strength – see *bond-dissociation energy.*

Bonding electron – see *valence electrons.*

Bonding molecular orbital – formed by the overlap of adjacent atomic orbitals.

Branched-chain *Alkane* – an *alkane* with *alkyl groups* bonded to the central carbon chain.

Brønsted-Lowry theory of acids and bases – a compound capable of donating a proton (a hydrogen ion); a Brønsted-Lowry base can accept a hydrogen ion; in *neutralization,* an acid donates a proton to a base, creating a conjugate acid and a conjugate base.

Buffer solution – a solution of definite pH made so that the pH alters only gradually by adding an acid or base.

C

Canonical structures – any of two or more hypothetical structures of *resonance* theory which can be written for a molecule simply by rearranging the valence electrons of the molecule (e.g., the two Kekule structures of benzene); sometimes called "valence bond isomers."

Carbanion – a carbon atom bearing a negative charge; a carbon anion.

Carbene – a reactive intermediate, characterized by a neutral, electron-deficient carbon center with two substituents - two single bonds and just six electrons in its valence shell ($R_2C:$).

Carbenoid – a chemical that resembles a carbene in its chemical reactions.

Carbocation – a carbon *Cation*; a carbon atom bearing a positive charge (sometimes referred to as a "carbonium ion").

Carbonyl group – a carbon double bonded to oxygen (C=O).

Carboxylic acid – a molecule containing a carboxyl (~COOH) group; also, a functional group.

$$-\!\!\overset{\displaystyle}{\underset{\displaystyle O}{C}}\!\!-OH$$

Catalyst – a substance that, when added to a reaction mixture, changes (speeds up) the rate of attainment of equilibrium in the system without itself undergoing a permanent chemical change.

Catalytic cracking – the method for producing gasoline from heavy petroleum distillates; generally, the catalysts are mixtures of silica and alumina or synthetic conjugates, such as zeolites.

Catalytic reforming – the process of improving the octane number of straight-run gasoline by increasing the proportion of *aromatic* and *branched-chain alkanes*; catalysts employed are either molybdenum-aluminum oxides or platinum-based.

Cation – a positively charged ion.

Cationic polymerization – occurs *via* a cation intermediate and is less efficient than *free-radical polymerization*.

Chain reaction – once started, produces enough energy to keep the reaction running; proceed by a series of steps which produce intermediates, energy, and products (e.g., the free radical addition of hydrogen bromide (HBr) to an *alkene)*.

Chair conformation – typically, the most stable cyclohexane conformation; looks like a chair.

Chemical shift – the location of an NMR peak relative to the standard tetramethylsilane (TMS), given in units of parts per million (ppm).

Chiral – describes a molecule that is not superimposable on its mirror image, like the relationship of a left hand to a right hand; see *chiral molecule.*

Chiral center – a carbon or other atom with four non-identical substituents.

Chiral molecule – a molecule that is not superimposable on its mirror image; rotate plane-polarized light.

Chirality – any asymmetric object or molecule; the property of non-identity of an object with its mirror image.

Chromatography – a series of related techniques for separating a mixture of compounds by their distribution between two phases; in gas-liquid chromatography, the distribution is between a gaseous and a liquid phase; in column chromatography, the distribution is between a liquid and a solid phase.

Cis – two identical substituents on the same side of a double bond or ring.

Closed-shell electron configuration – a stable electron configuration in which the electrons are in the lowest energy orbitals available.

Competing reactions – two reactions that start with the same reactants but form different products.

Compound – a term used generally to indicate a definite combination of elements into a more complex structure (a molecule); also applied to systems with non-stoichiometric proportions of elements.

Concerted – taking place at the same time without forming an intermediate.

Condensation reaction – a reaction in which two molecules join with the liberation of a small stable molecule.

Configuration – the order and relative three-dimensional orientation of the atoms in a molecule; given the designation R or S; "absolute configuration" is when the relative three-dimensional arrangement in the space of atoms in a *chiral* molecule has been correlated with an absolute standard.

Configurational isomers – a series of compounds that have the same constitution and bonding of atoms, but which differ in their atomic spatial arrangement (e.g., glucose and mannose).

Conformation – the spatial arrangement of a molecule in space at any moment; most molecules can adopt an infinite number of conformations because of rotation about single covalent bonds; of these possibilities, most compounds spend the most time in one or a few *preferred conformations*.

Conformer – a conformation of a molecule; generally, these will be at energy minima.

Conjugate acid – the acid that results when a Brønsted-Lowry base accepts a hydrogen ion.

Conjugate base – the base that results when a Brønsted-Lowry acid loses a hydrogen ion.

Conjugated double bonds – double bonds separated by a carbon-carbon single bond; alternating double bonds.

Conjugation – the overlapping in all directions of a series of p orbitals; a sequence of alternating double (or triple) and single bonds (e.g., C=C–C=C and C=C–C=O); can be relayed by the participation of lone pairs of electrons or vacant orbitals.

Conjugation energy – see *resonance energy*.

Constitution – the number and type of atoms in a molecule.

Constitutional isomers – molecules with the same molecular formula but with atoms attached differently.

Coordinate bond – the linkage of two atoms by a pair of electrons, both electrons provided by one of the atoms (the donor); covalent bonds.

Coupling constant (J) – the distance between two neighboring lines in an NMR peak (given in units of Hz); the separation in frequency units between multiple peaks in one chemical shift; this separation results from the spin-spin coupling.

Coupling protons – protons interacting and splitting the NMR peak into some lines following the $n+1$ rule.

Covalent bond – a bond formed by the sharing of electrons between atoms.

Cyano group – the ~C≡N group.

Cyanohydrin – a compound with the general formula:

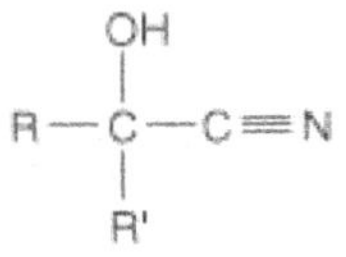

$$\begin{array}{c} OH \\ | \\ R-C-C\equiv N \\ | \\ R' \end{array}$$

Cyclization – the forming of ring structures.

Cycloaddition – a reaction that forms a ring.

Cycloalkane – a ring hydrocarbon made of carbon and hydrogen atoms joined by single bonds.

Cycloalkyl *Alkane* – an *alkane* to which a ring structure is bonded.

Cyclohydrocarbon – an *alkane*, *alkene* or *alkyne* formed in a ring structure rather than a straight or branched chain; the general formula is C_nH_{2n} (*n* must be a whole number of 3 or greater).

D

Deactivating group – causes an *Aromatic* ring to become less reactive toward electrophilic aromatic substitution.

Debye unit (D) – the unit of measure for a dipole moment; one debye equals 1.0×10^{-18} electrostatic units (esu · cm); see *dipole moment*.

Decarboxylation – a reaction in which carbon dioxide is expelled from a carboxylic acid.

Dehalogenation – the elimination reaction in which two halogen atoms are removed from adjacent carbon atoms to form a double bond.

Dehydration – the elimination reaction in which water is removed from a molecule.

Dehydrohalogenation – the elimination reaction in which a hydrogen atom and a halogen atom (a hydrohalic acid, like HBr, HCl) are removed from a molecule to form a double bond.

Delocalization – electron systems in which bonding electrons are not localized between two atoms as for a single bond but are spread (delocalized) over the whole group (e.g., *pi-bond* electrons the delocalized pi-electrons associated with *aromatic* molecules).

Delocalization energy – see *resonance energy*.

Delta value (δ value) – the chemical shift; the location of an NMR peak relative to the reference standard tetramethylsilane (TMS), given in units of parts per million (ppm).

Deprotonation – the loss of a proton (hydrogen ion) from a molecule.

Deshielding – an effect in NMR spectroscopy that the movement of *sigma* and *pi* electrons within the molecule causes; causes chemical shifts to appear at lower magnetic fields (downfield).

Dextrorotatory – the phenomenon in which plane-polarized light is turned in a clockwise direction.

Diastereomers (diastereoisomers) – stereoisomeric structures which are not enantiomers (mirror images); often applied to systems that differ in the configuration at one carbon (e.g., *meso-* and *d-* or *l-*tartaric acids).

Diels-Alder reaction – a cycloaddition between a conjugated diene and an *alkene* that produces a 1,4-addition product.

Diene – a molecule that contains two alternating double bonds; also, a reactant in the *Diels-Alder reaction*.

Dienophile – a reactant in the *Diels-Alder reaction* that contains a double bond; often substituted with electron-withdrawing groups.

Dienophile – the *alkene* that adds to the *diene* in a Diels-Alder reaction.

Dihalide – a compound that contains two halogen atoms; also called *"a dihaloalkane."*

Dihedral angle – the angle between groups attached on adjacent carbons when viewed in a *Newman projection*.

Diol – a compound that contains two hydroxyls (~OH) groups; also called a *"dihydroxy alkane."*

Dipole moment – a measure of the polarity of a molecule; the mathematical product of the charge in electrostatic units (ESU), and the distance that separates the two charges in centimeters (cm) (e.g., substituted *Alkyne*s have dipole moments caused by differences in electronegativity between the triple-bonded and single-bonded carbon atoms).

Disproportionation – a process in which a compound of one oxidation state changes to compounds of two or more oxidation states (e.g., $2 \, Cu^+ \rightarrow Cu + Cu^{2+}$).

Dissociation – the process whereby a molecule splits into simpler fragments that may be smaller molecules, atoms, free radicals, or ions.

Dissociation constant – the measure of the extent of dissociation, measured by the dissociation constant K. For the process:

$$AB = A + B$$

$$K = ([A] \cdot [B]) / [AB]$$

Dissymmetric – see *chiral*.

Distillation – the separation of components of a liquid mixture based on differences in boiling points.

Double bond – some atoms can share two pairs of electrons to form a double bond (two covalent bonds); formally, the second (double) bond arises from the overlap of *p* orbitals from two atoms, already united by a *sigma* bond, to form a *pi bond*; hydrocarbons that contain one double bond are *alkenes,* and hydrocarbons with two double bonds are *dienes.*

Doublet – describes an NMR signal split into two peaks.

Dyestuffs – intensely colored compounds applied to a substrate; colors are due to the absorption of light to give electronic transitions.

E

E₁ elimination reaction – a reaction that eliminates a hydrohalic acid (e.g., HCl, HBr) to form an *alkene*; a first-order reaction that goes through a *carbocation* mechanism.

E₂ elimination reaction – a reaction that eliminates a hydrohalic acid (e.g., HCl, HBr) to form an *alkene*; a second-order reaction that occurs in a single step in which the double bond is formed as the hydrohalic acid is eliminated.

E **isomer** – *stereoisomer* with the two highest priority groups on opposite sides of a ring or double bond.

Eclipsed – a *conformation* in which substituents on two attached saturated carbon atoms overlap when viewed as a *Newman projection*.

Eclipsed conformation – *conformation* about a carbon-carbon single bond in which the bonds off two adjacent carbons are aligned (0° apart when viewed in a *Newman projection*).

Electron – negatively charged particles of little weight that exist in quantized probability areas around the atomic nucleus.

Electron affinity – the amount of energy liberated when an electron is added to an atom in the gaseous state.

Electronegativity – measures an atom's ability to attract electrons toward itself in a covalent bond; the halogen fluorine is the most electronegative element; measured by the highest occupied molecular orbital (*HOMO*) and the lowest unoccupied molecular orbital (*LUMO*) energy levels.

Electronegativity scale – an arbitrary reference by which the electronegativity of elements can be compared.

Electronic configuration – the order in which electrons are arranged in an atom or molecule; used in a distinct and different sense from stereochemical *configuration*; see *stereochemistry*.

Electronic transition – in an atom or molecule, the electrons only have specific allowed energies (orbitals); if an electron passes from one orbital to another, an electronic transition occurs, and the emission or absorption of energy corresponds to the difference in energy of the two orbitals.

Electrophile – an "electron seeker;" an atom, molecule, or ion able to accept an electron pair to stabilize itself; a *Lewis acid*.

Electrophilic addition – a reaction in which the addition of an *electrophile* to an unsaturated molecule forms a saturated molecule.

Electrophilic substitution – an overall reaction in which an *electrophile* binds to a substrate with the expulsion of another electrophile (e.g., the electrophilic substitution of a proton by another electrophile, such as a nitronium ion, on an *Aromatic* substrate, such as benzene).

Electrostatic attraction – the attraction of a positive ion for a negative ion.

Electrovalent (ionic) bond – bonding by *electrostatic attraction*.

Element – a substance which cannot be further subdivided by chemical methods.

Element of unsaturation – a *pi bond*; a multiple bond or ring in a molecule.

Enantiomers – a pair of isomers related as mirror images of one another (e.g., isomers differing only in the configuration of the *chiral* atoms).

Enantiomorphic pair – in optically active molecules with more than one stereogenic center, the two structures are mirror images.

Endothermic – a reaction in which heat is absorbed.

Energy diagram (or *reaction energy diagram*) – a graph of the energy against the progress of the reaction.

Energy of reaction – the difference between the total energy content of the reactants and the total energy content of the products; the greater the energy of reaction, the more stable the products.

Enol – an unstable compound (e.g., *vinyl alcohol*) in which a hydroxide group is attached to a carbon in a carbon-carbon double bond; these compounds *tautomerize* to form more stable *ketones*.

Enolate ion – the resonance stabilized ion formed when an *aldehyde* or *ketone* loses an α hydrogen:

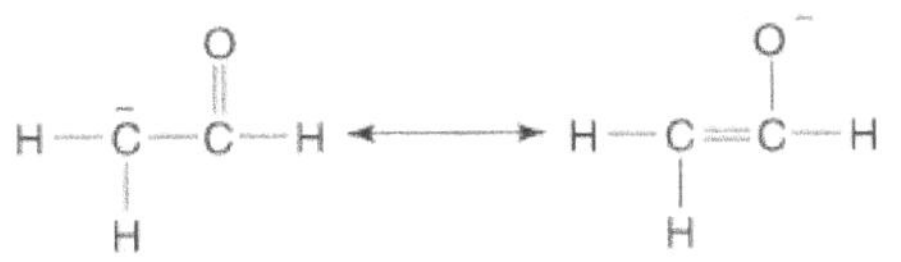

Enthalpy (*H*) – a thermodynamic state function, generally measured in kilojoules per mole; in chemical reactions, the enthalpy change (ΔH) is related to changes in the free energy (ΔG) and *entropy* (ΔS) by the equation: $\Delta G = \Delta H - T\Delta S$.

Entropy (*S*) – a thermodynamic quantity that is a measure of the degree of disorder within a system. The greater the degree of order, the higher the entropy; for an increase in entropy, S is positive; has the units of joules per degree K per mole.

Enzyme – a naturally occurring substance able to catalyze a chemical reaction.

Epimerization – a process in which the configuration about one *chiral* center of a compound, containing more than one *chiral* atom, is inverted to give the opposite configuration; the term "epimers" describes two related compounds that differ only in the configuration about one chiral atom.

Epoxide – a three-membered ring that contains oxygen.

Epoxidation – the addition of an oxygen bridge across a double bond to give an oxirane; achieved by use of a peracid or, in a few cases, by use of a *Catalyst* and oxygen.

Equatorial – the bonds in a chair cyclohexane oriented along the equator of the ring.

Equilibrium constant – according to the law of mass action, for a reversible chemical reaction, aA + bB = cC + dD, the equilibrium constant (K) is defined as: $K = ([C]^c[D]^d) / ([A]^a[B]^b)$.

Ester – a functional group; a molecule containing a carbonyl group adjacent to an oxygen (RCOOR').

$$-\overset{\overset{\textstyle O}{\|}}{C}-OR$$

Ether – a molecule containing oxygen singly-bonded to two carbon atoms; also, a functional group; the general formula is R—O—R'; epoxyethane, an epoxide, is a cyclic ether; often refers to diethyl ether.

Excited state – the state of an atom, molecule, or group when it has absorbed energy and becomes excited to a higher energy state than the ground state; may be electronic, vibrational, rotational, etc.

F

Fischer projection – a convention for drawing carbon chains so that the relative three-dimensional stereochemistry of the carbon atoms is relatively easy to portray as a 2-dimensional drawing.

Fingerprint region – an IR spectrum below 1,500 cm^{-1}; often complex and difficult to interpret.

Free energy (ΔG) – a thermodynamic state function; the free energy change (ΔG) in any reaction is related to the *enthalpy* and *entropy*: $\Delta G = \Delta H - T\Delta S$.

Free radicals – molecules or ions with unpaired electrons; generally, extremely reactive; "stable" free radicals include molecular oxygen, NO, and NO_2; organic free radicals range from those of transient existence only to very long-lived species; alkyl free radicals tend to be very reactive and short-lived.

Free-radical chain reaction – a reaction that proceeds by a free-radical intermediate in a chain mechanism (a series of self-propagating, interconnected steps); compare with a *free-radical reaction*.

Free-radical polymerization – a polymerization initiated by a *free radical*.

Free-radical reaction – a reaction in which a covalent bond is formed by the union of two radicals; compare with a *free-radical chain reaction*.

Frontier orbital symmetry – the theory that the site and rates of reaction depend on the geometries, the sign of the wave function and relative energies of the highest occupied molecular orbital (*HOMO*) of one molecule and the lowest unoccupied molecular orbital (*LUMO*) of the other.

Functional group – a set of bonded atoms that displays a specific molecular structure and chemical reactivity when bonded to a carbon atom in the place of a hydrogen atom.

G

Gauche – a conformational isomer in which the groups are neither eclipsed nor trans to one another; often taken as the conformation where the dihedral angle between the groups is 60°.

Gauche conformation – a type of staggered conformation in which two bulky groups are next to each other.

Geometrical isomerism – isomerism from the restricted rotation about a bond (e.g., (*Z*) and (*E*) isomers of unsymmetrically substituted *alkene*s).

Grignard reagent – an organometallic reagent in which magnesium metal inserts between an *alkyl group* and a halogen (e.g., CH_3MgBr).

Ground state – the lowest energy state of an atom, molecule, or ion.

H

Half-life, $t^{1/2}$ – the time taken for the concentration of a substance in a reaction to reduce to half its original value; used in first-order reactions and as a measure of the rate of radioactive decay.

Halide – a member of the VIIA column of the periodic table (e. g., F, Cl, Br, I) or a molecule that contains one of these atoms; also, a functional group.

Haloalkane – an *alkane* that contains one or more halogen atoms; also called an *alkyl halide*.

Halogen – an electronegative, nonmetallic element in Group VII of the periodic table, including fluorine, chlorine, bromine, and iodine; often represented in structural formulas by an "X."

Halogenation – a reaction in which halogen atoms are bonded to an *Alkene* at the double bond.

Halonium ion – a halogen atom that bears a positive charge; highly unstable.

Hard and soft acids and bases – a classification of acids and bases depending on their polarizability; hard bases include fluoride ions; soft bases include triphenylphosphine; hard acids include Na^+, whilst an example of a soft, polarizable acid is Pt^{2+}; hard-hard and soft-soft interactions are favored; hardness and softness can be described in terms of the *HOMO* and *LUMO* interactions.

Heat of reaction – the amount of heat absorbed or evolved when specified amounts of compounds react under constant pressure; expressed as kilojoules per mole; for exothermic reactions, the convention is that the *Enthalpy* (*H*) change (heat of reaction) is negative.

Hemiacetal – a functional group with a hydroxyl and ether attached to the same carbon; the structure:

Hemiketal – a functional group with a hydroxyl and ether attached to the same carbon; the structure:

$$-\overset{\displaystyle |}{\underset{\displaystyle R}{C}}\Big\langle\begin{smallmatrix}OR\\OH\end{smallmatrix}$$

Hertz – a measure of a wave's frequency; equals the number of waves that passes a specific point per second.

Heteroatom – in organic chemistry, an atom other than carbon.

Heterocyclic compound – a class of cyclic compounds in which one of the ring atoms is not carbon (e.g., epoxyethane).

Heterogeneous reaction – occurs between substances mainly present in different phases (e.g., between a gas and a liquid).

Heterogenic bond formation – a type of bond formed by the overlap of orbitals on adjacent atoms. One orbital of the pair donates both electrons to the bond.

Heterolytic cleavage – the fracture of a bond so that one of the atoms receives both electrons; in reactions, this asymmetrical bond rupture generates carbocation and carbanion mechanism.

Heterolytic reaction – a reaction with a covalent bond broken by unequal sharing of bonding electrons.

HOMO – the highest occupied molecular orbital of a molecule, ion, or atom.

Homologous series – compounds with common compositions (e.g., *alkane*s, *alkene*s and *alkyne*s).

Homolog – one of a series of compounds in which each member differs by a constant unit.

Homolytic cleavage – the fracture of a bond in such a manner that both atoms receive one of the bond's electrons; this symmetrical bond rupture forms free radicals; in reactions, it generates *free-radical* mechanisms.

Homolytic reaction – a covalent bond is broken with equal sharing of the electrons from the bond.

Hückel's rule – a compound with $4n + 2$ π electrons has a closed-shell electron configuration and is *aromatic*.

Hybrid orbitals – formed from mixing atomic orbitals (AO), like the sp^x orbitals, which result from mixing s and p orbitals.

Hydration – the addition of the elements of water to a molecule.

Hydride shift – the movement of a hydride ion (a hydrogen atom with a negative charge; H^-) to form a more inductively-stabilized carbocation.

Hybridization – the process whereby atomic orbitals of different types but similar energies are combined to form a set of equivalent hybrid orbitals; these hybrid orbitals do not exist in the atoms but only by forming molecular orbitals by combining atomic orbitals from different atoms.

Hydroboration – the *cis*-addition of B–H bonds across double (or triple) carbon-carbon bonds.

Hydroboration-oxidation – the addition of borane (BH_3) or an alkyl borane to an *alkene* and its subsequent oxidation to produce the *anti-Markovnikov* indirect addition of water.

Hydrocarbon – a molecule that exclusively contains carbon and hydrogen atoms; the central bond may be a single, double, or triple covalent bond, and it forms the molecule's backbone.

Hydrogenation – the addition of hydrogen to a multiple bond.

Hydrogenolysis – the cleaving of a chemical bond by hydrogen, generally with a hydrogenation *catalyst*.

Hydrohalogenation – a reaction in which a hydrogen atom and a halogen atom are added to a double bond to form a saturated compound.

Hydrolysis – the addition water to a substance, often with the partition of the substance into two parts (e.g., the hydrolysis of an *ester* to an acid and an *alcohol*).

Hydrolyze – to cleave a bond *via* the elements of water.

Hyperconjugation – weak interaction (electron donation) between *sigma* bonds with *p* orbitals; explains why alkyl substituents stabilize carbocations.

I

Inductive effect – an electronic effect transmitted through bonds in an organic compound due to the electronegativity of substituents and the permanent polarization thereof; the substituent either induces charges towards or away from itself with the formation of a dipole.

Infrared spectroscopy (IR) – the study of the absorption of infrared light by substances; corresponding to vibrational (and some rotational) changes, infrared spectroscopy provides valuable information about the molecule's structure; detailed correlation tables exist relating infrared bands (absorbances) to functional groups.

Inhibitor – a general term for a compound that inhibits (slows down) a reaction; can be used to slow or stop free radical chain reactions.

Initiation step – the first step in the mechanism of a reaction.

Initiator – a material capable of being fragmented into free radicals, which initiates a *free-radical reaction*.

Insertion – placing between two atoms.

Intermediate – a species that form in one step of a multistep mechanism; unstable and cannot be isolated.

Ion – an atom or group of atoms that has lost or gained one or more electrons to become a charged species.

Ionic bond – a bond formed by the transfer of electrons between atoms, resulting in forming ions of opposite charge; the electrostatic attraction between these ions.

Ionization energy – the energy needed to remove an electron from an atom.

IR spectroscopy – an instrumental technique that measures IR (infrared) light absorption by molecules and can determine functional groups in an unknown molecule.

Isolated double bond – a double bond more than one single bond away from another double bond in a *diene*.

Isomers – compounds having the same atomic composition (*constitution*) but differing in their chemical structure; includes structural isomers (chain or positional), tautomeric isomers, and stereoisomers (including geometrical isomers, optical isomers, and conformational isomers).

IUPAC nomenclature – a systematic method for naming molecules based on a series of rules developed by the International Union of Pure and Applied Chemistry, not the only reference body, but the most common.

J

J **value** – the coupling constant between two peaks in an NMR signal; given in units of Hz.

K

Kekulé structure – the structure for benzene in which there are three alternating double and single bonds in a six-membered ring of carbon atoms.

Ketal – the product formed by the reaction of a *ketone* with *alcohol*; the general structure is:

$$R-C\begin{smallmatrix} OR' \\ \\ | \\ R \end{smallmatrix}OR'$$

Keto-enol tautomerization – the process by which an *enol* equilibrates with its corresponding *aldehyde* or *ketone*.

Ketone – a compound in which an oxygen atom is bonded *via* a double bond to a carbon atom, bonded to two more carbon atoms.

Kinetic product – the product that forms the fastest; has the lowest *activation energy*.

Kinetics – the study of the rate of reactions.

Kinetically controlled product – the product formed from the fastest reaction in competing reaction pathways.

L

Levorotatory – the phenomenon that turns plane-polarized light in a counterclockwise direction.

LCAO – a method for calculating molecular orbitals from a "Linear Combination of Atomic Orbitals."

Leaving group – the negatively charged group that departs from a molecule undergoing a *nucleophilic substitution* reaction.

Lewis acid – an agent capable of accepting a pair of electrons to form a *coordinate bond*.

Lewis base – an agent capable of donating a pair of electrons to form a *coordinate bond*.

Lone pair – a pair of electrons in a molecule not shared by two of the constituent atoms.

Linear – the shape of a molecule with *sp* hybrid orbitals; an *alkyne*.

LUMO – the lowest unoccupied molecular orbital in a molecule or ion.

M

Markovnikov rule – the positive part of a reagent (e.g., a hydrogen atom) adds to the carbon of the double bond that already has more hydrogen atoms attached; the negative part adds to the other carbon of the double bond; leading to the more stable *carbocation* over other less-stable intermediates; useful to predict the major product; *free radical reactions* proceed in the opposite sense, giving rise to *anti-Markovnikov addition*.

Mass number – the total number of protons and neutrons in an atom.

Mass spectrometry – a form of spectrometry in which, generally, high energy electrons are bombarded onto a sample, generating charged fragments of the parent substance; electrostatic and magnetic fields then focus these ions for a spectrum of the charged fragments.

Mechanism – the series of steps that reactants go through during their conversion into products.

Meso compounds – molecules with *chiral* centers but are *achiral* due to one or more planes of symmetry.

Mesomerism – see *resonance*.

Meta – describes the positions of two substituents on a *benzene ring* separated by one carbon.

Meta-directing substituent – groups on an *aromatic* ring directing incoming *electrophile*s to meta position.

Methylene group – a ~CH_2~ group.

Microwave spectroscopy – the interaction of electromagnetic waves with wavelengths in the range 10^{-2} to 1 meter; this energy range corresponds to rotational frequencies; helpful in studying the structure of materials (generally gases) and their characterization.

Molecular ion – the fragment in a mass spectrum corresponding to the *cation* radical (M+) of the molecule; gives the molecular mass of the molecule.

Molecular orbitals – the electron orbitals belonging to a group of atoms forming a molecule.

Molecular orbital theory – a model for depicting the location of electrons that allows electrons to delocalize across the entire molecule, a more accurate but less user-friendly theory than the *valence bond theory*.

Molecule – a covalently-bonded collection of atoms with no electrostatic charge; the smallest particle of matter that can exist in a free state; in the case of ionic substances, such as sodium chloride, the molecule is considered as a pair of ions (e.g., NaCl).

Multiple bonds – a double or triple bond; atomic *p* orbitals in side-to-side overlap, preventing rotation.

Multistep synthesis – synthesis of a compound that takes several steps to achieve.

N

n+1 rule – rule for predicting the coupling for a proton in ^{1}H NMR spectroscopy; an NMR signal splits into n+1 peak, where n is the number of equivalent adjacent protons.

Natural product – a compound produced by a living organism.

Neutralization – the reaction of an acid and a base; the acid and base reaction products are salt and water.

Neutron – an uncharged particle in the atomic nucleus with the same weight as a proton; additional neutrons do not change an element but convert it to one of its isotopic forms.

Newman projection – a projection obtained by viewing along a carbon-carbon single (double) bond.

Nitrile – a compound with a cyano group (a carbon triply-bonded to nitrogen ($\sim$C$\equiv$N)); a functional group.

NMR – nuclear magnetic resonance spectroscopy; a technique that measures radiofrequency light absorption by molecules; a powerful structure-determining method; see *nuclear magnetic resonance spectroscopy*.

Node – a region of zero electron density in an orbital; a point of zero amplitude in a wave.

Nonbenzenoid aromatic ring – an *aromatic* ring system that does not contain a *Benzene ring*.

Nonbonding electrons – *valence electrons* not used for covalent bond formation.

Nonterminal Alkyne – an *alkyne* in which the triple bond is somewhere other than the 1 position.

Nuclear magnetic resonance (NMR) spectroscopy – a form of spectroscopy that depends on the absorption and emission of energy arising from changes in the spin states of the nucleus of an atom; for aggregates of atoms, as in molecules, minor variations in these energy changes are caused by the local chemical environment; the energy changes used are in the radiofrequency range of the electromagnetic spectrum and depend upon the magnitude of an applied magnetic field.

Nucleofuge – see *leaving group*.

Nucleophile – a "nucleus lover;" a molecule with the ability to donate a lone pair of electrons (a *Lewis base*).

Nucleophilic substitution – an overall reaction in which a *nucleophile* reacts with a compound displacing another nucleophile; such reactions commonly occur in aliphatic chemistry; if the reaction is unimolecular, they are *S_N1 reactions*; for bimolecular reactions, they are *S_N2 reactions*.

Nucleophilicity – a measure of the reactivity of a *nucleophile* in a *nucleophilic substitution* reaction.

Nucleus – the central core of an atom; the location of the protons and neutrons.

O

Optical activity – the property of certain substances to rotate plane-polarized light; associated with asymmetry; compounds that possess a *chiral* carbon atom of the same "handedness" rotate plane-polarized light; isomers that rotate light in equal but opposite directions are "optical isomers," although the better term is *enantiomers*.

Orbit – an area around an atomic nucleus with a high probability of finding an electron; also called a *shell*, divided into *orbitals* or *subshells*.

Orbital – an area in an *orbit* with a high probability of finding an electron; a "subshell;" the orbitals have the same principal and angular quantum numbers.

Organic compound – carbon-containing compound.

Ortho – describes the positions of two substituents on a *benzene ring* on adjacent carbons.

Ortho-para director – an *aromatic* substituent that directs incoming *electrophile*s to *ortho* or *para* positions.

Outer-shell electron – see *valence electrons*.

Overlap region – the region in space where atomic or molecular orbitals overlap, creating an area of high electron density.

Oxidation – a chemical process in which the proportion of electronegative substituents in a compound is increased (the loss of electrons by an atom in a covalent bond), the charge is made more positive, or the oxidation number is increased; in organic reactions, when a compound accepts additional oxygen atoms.

Oxonium ion – a positively-charged oxygen atom.

Ozonide – a compound formed by the addition of ozone to a double bond.

Ozonolysis – the cleavage of double and triple bonds by ozone, O_3.

P

Paired spin – the spinning in opposite directions of the two electrons in a bonding orbital.

Para – describes the positions of two substituents on a *benzene ring* separated by two carbons.

Parent name – the root name of a molecule according to the *IUPAC nomenclature* rules (e.g., hexane is the parent name in *trans*-1,2-dibromocyclohexane).

Peroxide – a compound that contains an oxygen-oxygen single covalent bond.

Peroxyacid – an acid of the general form:

$$R-\overset{\overset{\textstyle O}{\|}}{C}-O-OH$$

Phenyl ring – a *benzene ring* as a substituent, abbreviated Ph.

Photochemical reaction – a chemical reaction brought about by the action of light.

Pi **(π) bond** – formed by the side-to-side overlap of atomic *p* orbitals (with electron density above and below the two atoms, but not directly between the two atoms); weaker than a *sigma bond* because of poor orbital overlap caused by nuclear repulsion; create unsaturated molecules; found in double and triple bonds.

Pi **(π) complex** – an intermediate formed when a *cation* is attracted to the high electron density of a *pi bond*.

Pi **(π) molecular orbital** – a molecular orbital created by the side-to-side overlap of atomic *p* orbitals.

p*K*a – the scale for defining a molecule's acidity (p*K*a = –log K_a).

Plane-polarized light – light that oscillates in a single plane.

Plane of symmetry – a plane cutting through a molecule in which both halves are mirror images of each other.

Polar covalent bond – a bond in which the shared electrons are not equally available in the overlap region, forming partially positive and partially negative ends on the molecule.

Polarimeter – measures rotation of plane-polarized light by a compound, generally prepared in a solution.

Polarity – the asymmetrical distribution of electrons in a molecule, positive and negative ends on the molecule.

Precursor – the substance from which another compound is formed.

Preparation – a reaction in which the desired chemical is produced (e.g., the dehydration of an *alcohol* is a preparation for an *alkene)*.

Primary carbocation – a *carbocation* to which one *alkyl group* is bonded.

Primary (1°) carbon – a carbon atom that is attached to one other carbon atom.

Product – the substance that forms when reactants combine in a reaction.

Propagation step – the event in a free radical reaction in which both a product and energy are produced; the energy keeps the reaction going.

Protecting group – formed on a molecule by the reaction of a reagent with a substituent on the molecule; the resulting group is less sensitive to further reaction than the original group, but it must be easy to be reconverted to the original group.

Protic solvent – a solvent that contains O–H or N–H bonds.

Proton – an H^+ ion; also, a positively-charged nuclear particle.

Protonation – the addition of a proton (a hydrogen ion) to a molecule.

Pure covalent bond – in which the shared electrons are equally available to the bonded atoms.

Pyrolysis – the application of high temperatures to a compound.

R

R group – abbreviation given to an unimportant part of a molecule; indicates rest of molecule; see *alkyl group*.

Racemate – a 50:50 mixture of two enantiomers; another name for *racemic mixture*.

Racemic mixture – an equimolar mixture of the two enantiomeric isomers of a compound; because of the equal numbers of *levo-* and *dextro*-rotatory molecules present in a *racemate*, there is no net rotation of plane-polarized light (i.e., they are optically inactive).

Radical – an atom or molecule having one or more free valences; see *free radicals*.

Rate-determining step – the step in a reaction's mechanism that requires the highest activation energy and is, therefore, the slowest.

Rate of reaction – the speed with which a reaction proceeds.

Reactant – a starting material.

Reaction energy – the difference between the energy of the reactants and that of the products.

Reagent – the chemicals that ordinarily produce reaction products.

Rearrangement reaction – a reaction that causes the skeletal structure of the reactant to change in converting to product.

Reduction – chemical processes in which the proportion of more electronegative substituents is decreased, the charge is made more negative, or the oxidation number is lowered.

Resolution – the separation of a *racemate* into its two enantiomers using some *chiral* agency.

Resonance – 1) the representation of a compound by two or more canonical structures in which the *valence electrons* are rearranged to give structures of similar probability; the actual structure is a hybrid of the resonance forms; 2) the process by which a substituent removes electrons from or gives electrons to a *pi bond* in a molecule; a delocalization of electrical charge in a molecule.

Resonance energy – the difference in energy between the calculated energy content of a *resonance structure* and the actual energy content of the hybrid structure.

Resonance hybrid – the actual structure of a molecule that shows resonance; possesses the characteristics of possible structures (and consequently cannot be drawn); lower in energy than any structure for the molecule and is more stable.

Resonance structures – intermediate structures of one molecule that differ only in the positions of their electrons; used to depict the location of *pi* and nonbonding electrons on a molecule; a molecule looks like a hybrid of all resonance structures; none of the drawn resonance structures are correct, and the best representation is a hybrid of the drawn structures.

Reversible process – the forward reaction can reach an equilibrium with the reverse reaction.

Ring structure – a molecule in which the end atoms have bonded, forming a ring rather than a straight chain.

Rotamers – isomers formed by restricted rotation.

Rotation – the ability of carbon atoms attached by single bonds to freely turn, which gives the molecule an infinite number of *conformations*.

R/S convention – a formal non-ambiguous nomenclature system for assigning absolute configuration of structure to *chiral* atoms using the Cahn, Ingold and Prelog priority rules.

S

s-*cis* conformation – a relationship in which the two double bonds of a conjugated *diene* are on the same side (*cis*) of the carbon-carbon single bond connecting them; the required conformation for the *Diels-Alder reaction*.

s-*trans* conformation – the *conformation* in which the two double bonds of a conjugated *diene* are on opposite (*trans*) sides of the carbon-carbon single bond that connects them.

Saturated – the term given to organic molecules that contain no multiple bonds.

Saturated compound – a compound containing single bonds.

Saturation – the condition of a molecule containing the most atoms possible; a molecule with single bonds.

Sawhorse projection – the sideways projection of a carbon-carbon single bond and the attached substituents; gives a clearer representation of stereochemistry than the *Fischer projection*; see *Newman projection*.

Secondary carbocation – a positively charged intermediate to which two *alkyl group*s are bonded.

Secondary (2°) carbon – a carbon atom that is directly attached to two other carbon atoms.

Separation technique – a process by which products are isolated from each other and impurities.

Shielding – an effect, in NMR spectroscopy, caused by the movement of *sigma* and *pi* electrons within the molecule; causes chemical shifts to appear at higher magnetic fields (upfield).

***Sigma* (σ) antibonding molecular orbital** – in which one or more electrons are less stable than when localized in the isolated atomic orbitals from which the molecular orbital (MO) was formed.

***Sigma* (σ) bond** – formed by the linear combination of orbitals so that the maximum electron density is along a line joining the two nuclei of the atoms.

***Sigma* (σ) bonding molecular orbital** – when electrons are more stable than localized in the isolated atomic orbitals from which the molecular orbital was formed.

Singlet – describes an NMR signal consisting of only one peak.

Skeletal structure – the carbon backbone of a molecule.

S$_N$1 – a *substitution reaction* mechanism in which the slow step is a self-ionization of a molecule to form a *carbocation*; the rate-controlling step is unimolecular.

S$_N$1 reaction – a first-order substitution reaction that goes through a *carbocation* intermediate.

S$_N$2 – a *substitution reaction* mechanism in which the rate-controlling step is a simultaneous attack by a *Nucleophile* and a departure of a *leaving group* from a molecule; the rate-controlling step is bimolecular.

S$_N$2 reaction – a second-order *substitution reaction* that takes place in one step and has no intermediates during the reaction pathway.

***sp* hybrid orbital** – a molecular orbital (MO) created by combining wave functions of an *s* and a *p* orbital.

***sp*^{2} hybrid orbital** – a molecular orbital (MO) created by combining wave functions of an *s* and two *p* orbitals.

***sp*^{3} hybrid orbital** – a molecular orbital (MO) created by combining wave functions of an *s* and three *p* orbitals.

Spectrometer – an instrument that measures the spectrum of a sample (e.g., a *mass spectrometer*).

Spectrophotometer – an instrument that measures the degree of absorption (or emission) of electromagnetic radiation by a substance. The measuring system generally includes a photomultiplier UV., IR, visible and microwave regions of the electromagnetic spectrum.

Spin-spin splitting – NMR (nuclear magnetic resonance) signals caused by the coupling of nuclear spins on neighboring nonequivalent hydrogens.

Stability constant – when a complex is formed between a metal ion and a ligand in solution, the equilibrium may be expressed by a constant related to the free energy change for the process: $M + A = MA : \Delta G = -RT\ln K$.

Staggered conformation – the orientation about a carbon-carbon single bond in which bonds off one carbon are at a maximum distance apart from bonds coming from an adjacent carbon (60° apart when viewed in a *Newman projection*).

Stereochemistry – the study of the spatial arrangements of atoms in molecules and complexes.

Stereoisomers – molecules that have the same atom connectivity but different orientations of those atoms in three-dimensional space. Another name for *configurational isomer*.

Stereospecific reactions – bonds are broken and made at a particular carbon atom and lead to a single stereoisomer; if the configuration is altered in the process, the reaction undergoes inversion of configuration; if the configuration remains the same, the transformation occurs with retention of configuration.

Steric hindrance – a physical blockage of a site within a molecule by the presence of local atoms or groups of atoms; therefore, a reaction at a particular site will be impeded.

Straight-chain alkane – a saturated hydrocarbon that has no carbon-containing side chains.

Structural isomer (or *constitutional isomer*) – have the same molecular formula but different bonding among their atoms (e.g., C_4H_{10} can be butane or 2-methylpropane, and C_4H_8 can be 1-butene or 2-butene).

Subatomic particles – a component of an atom; a proton, neutron, or electron.

Substituent – a piece that sticks off the main carbon chain or ring.

Substituent group – an atom or group that replaces a hydrogen atom on a hydrocarbon.

Substitution – the replacement of an atom or group bonded to a carbon atom with a second atom or group.

Substitution reactions – an atom or group of atoms replace one atom or group of atoms; see *electrophilic substitutions* and *nucleophilic substitutions*.

Syn addition – a reaction in which two groups of a reagent X–Y add on the same face of a carbon-carbon double bond.

T

Tautomerism – a form of structural isomerism where the two structures are interconvertible using the migration of a proton.

Tautomers – molecules that differ in the placement of hydrogen and double bonds and are easily interconvertible; keto and *enol* forms are tautomers; see *keto-enol tautomerization*.

Terminal alkyne – an *alkyne* whose triple bond is between the first and second carbon atoms of the chain.

Terminal carbon – the carbon atom on the end of a carbon chain.

Termination step – the step in a reaction mechanism ending the reaction, often between two free radicals.

Tertiary carbocation – a *carbocation* to which three *alkyl group*s are bonded.

Tertiary (3°) carbon – a carbon atom that is directly attached to three other carbon atoms.

Tetrahaloalkane – an *alkane* that contains four halogen atoms on the carbon chain; the halogen atoms can be on vicinal or non-vicinal carbon atoms.

Thermodynamic product – the reaction product with the lowest energy.

Thermodynamically controlled reaction – conditions permit two or more products to form; the products are in an equilibrium condition, allowing the more stable product to predominate.

Thermodynamics – the study of the energies of molecules.

Thiol – a molecule containing an –SH group; also, a functional group.

Tosyl group – a *p*-toluenesulfonate group:

Tosylation – a reaction that introduces the toluene-4-sulphonyl group into a molecule, generally by reacting an *alcohol* with tosyl chloride to give the tosylate *ester*.

Transition state – the point of highest energy on an energy against reaction coordinate curve; the least stable point (peak) on a reaction path; a reaction path may involve more than one transition state.

Trigonal planar – the shape of a molecule with an sp^2 hybrid orbital; in this arrangement, the *sigma bonds* are in a single plane separated by 60° angles.

Triple bond – a multiple bond composed of one *sigma bond* and two *pi bonds*; rotation is not possible around a triple bond; hydrocarbons containing triple bonds are *alkynes*.

Triplet – describes an NMR signal split into three peaks.

U

Ultraviolet light (UV) – radiation of a higher energy range than visible light but lower than that of ionizing radiations (e.g., X-rays); many substances absorb ultraviolet light, leading to electronic excitation; useful both for characterizing materials and stimulating chemical reactions (*photochemical reactions*).

Ultraviolet spectroscopy – spectroscopy that measures how much energy a molecule absorbs in the ultraviolet region of the spectrum.

Unsaturated – an organic compound containing multiple bonds.

Unsaturated compound – contain one or more multiple bonds (e.g., *alkene*s and *alkyne*s).

Unsaturation – a molecule containing less than the maximum number of single bonds due to multiple bonds.

V

Valence bond theory – the mechanical wave basis of *resonance* theory.

Valence electrons – the outermost electrons of an atom (e.g., the valence electrons of the carbon atom occupy the $2s$, $2p_x$ and $2p_y$ orbitals).

Valence isomerization – the isomerization of molecules that involve structural changes resulting only from a relocation of single and double bonds; if a dynamic equilibrium is established between the two isomers, it is referred to as "valence *tautomerism*" (e.g., the valence tautomerism of cyclo-octa-1,3,5-triene).

Valence shell – the outermost electron orbit.

Vinyl alcohol – ~CH_2=CH–OH

Vinyl group – the ethenyl group: ~CH=CH_2.

W

Walden inversion – a Walden inversion occurs at a tetrahedral carbon atom during an S_N2 *reaction* when the entry of the reagent and the departure of the leaving group are synchronous; the result is an inversion of configuration at the center under attack.

Wurtz reaction – the coupling of two *alkyl halide* molecules to form an *alkane*.

X

X group – "X" is the abbreviation for a halogen substituent in the structural formula of an organic molecule.

Y

Ylide – a neutral molecule in which two oppositely charged atoms are bonded to each other.

Z

Z isomer – the two highest-priority substituents are on the same side of a double bond or ring.

Zaitsev's rule – the major product in the formation of *alkene*s by elimination reactions will be the more highly substituted alkene or the alkene with more substituents on the carbon atoms of the double bond.

Isomer Classification

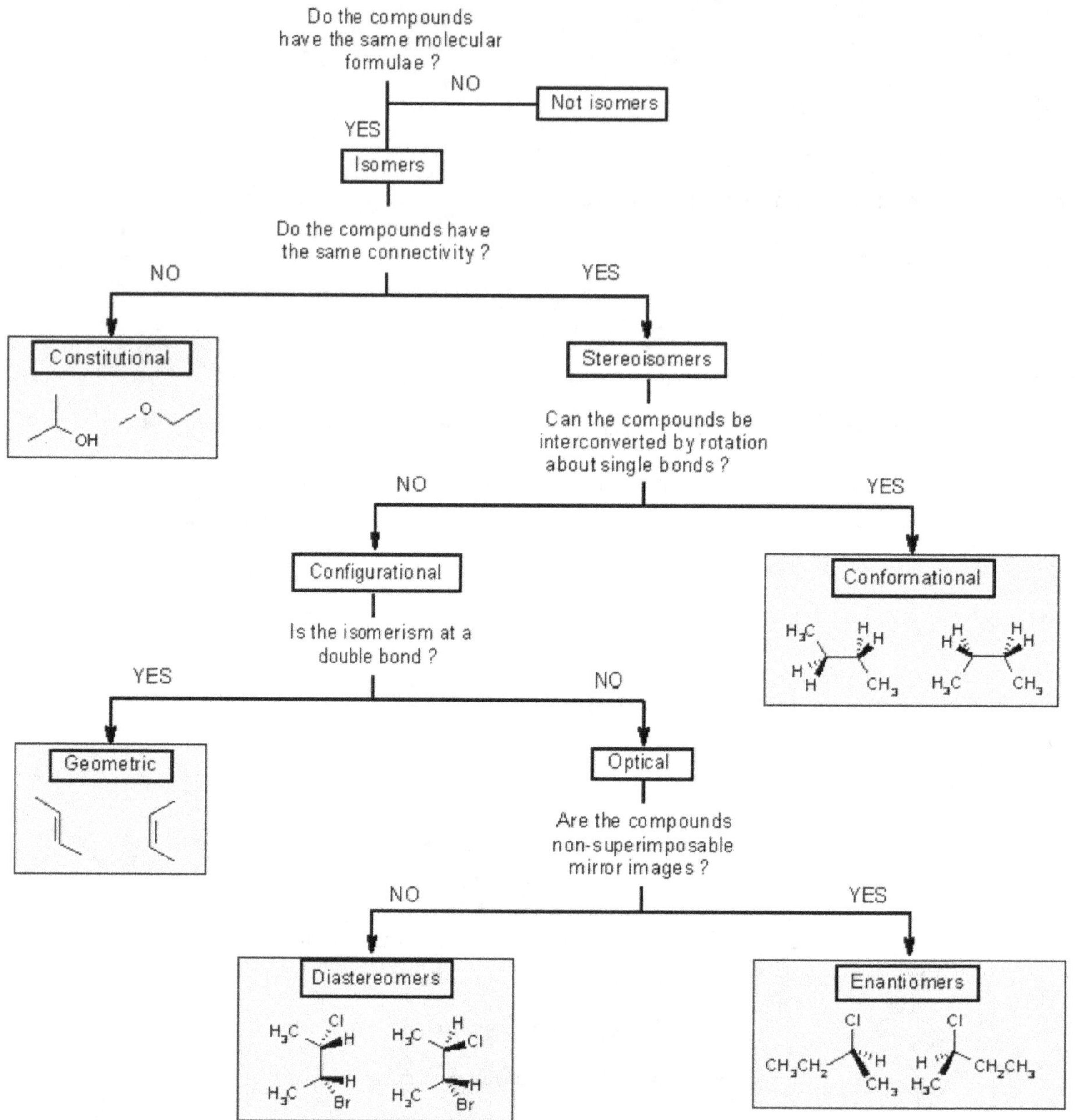

Periodic Table of the Elements

Key: Atomic Number · Valence · **Symbol** · Name · Atomic Mass

Main Group and Transition Elements

Group	1 IA 1A	2 IIA 2A	3 IIIB 3B	4 IVB 4B	5 VB 5B	6 VIB 6B	7 VIIB 7B	8 VIII	9 VIII	10 VIII	11 IB 1B	12 IIB 2B	13 IIIA 3A	14 IVA 4A	15 VA 5A	16 VIA 6A	17 VIIA 7A	18 VIIIA 8A
1	1 **H** Hydrogen 1.008 (+1,-1)																	2 **He** Helium 4.003 (0)
2	3 **Li** Lithium 6.941 (+1)	4 **Be** Beryllium 9.012 (+2)											5 **B** Boron 10.811 (+3)	6 **C** Carbon 12.011 (+4,+3,+2,+1,-4)	7 **N** Nitrogen 14.007 (+5,+3,-3)	8 **O** Oxygen 15.999 (-2)	9 **F** Fluorine 18.998 (-1)	10 **Ne** Neon 20.180 (0)
3	11 **Na** Sodium 22.990 (+1)	12 **Mg** Magnesium 24.305 (+2)											13 **Al** Aluminum 26.982 (+3)	14 **Si** Silicon 28.086 (+4,-4)	15 **P** Phosphorus 30.974 (+5,+3,-3)	16 **S** Sulfur 32.066 (+6,+4,+2,-2)	17 **Cl** Chlorine 35.453 (+7,+5,+3,+1,-1)	18 **Ar** Argon 39.948 (0)
4	19 **K** Potassium 39.098 (+1)	20 **Ca** Calcium 40.078 (+2)	21 **Sc** Scandium 44.956 (+3)	22 **Ti** Titanium 47.88 (+4)	23 **V** Vanadium 50.942 (+5)	24 **Cr** Chromium 51.996 (+6,+3)	25 **Mn** Manganese 54.938 (+7,+4,+2)	26 **Fe** Iron 55.845 (+6,+3,+2)	27 **Co** Cobalt 58.933 (+3,+2)	28 **Ni** Nickel 58.693 (+2)	29 **Cu** Copper 63.546 (+2,+1)	30 **Zn** Zinc 65.38 (+2)	31 **Ga** Gallium 69.723 (+3)	32 **Ge** Germanium 72.631 (+4,+2,-4)	33 **As** Arsenic 74.922 (+5,+3,-3)	34 **Se** Selenium 78.971 (+6,+4,-2)	35 **Br** Bromine 79.904 (+5,+3,+1,-1)	36 **Kr** Krypton 84.798 (+2,0)
5	37 **Rb** Rubidium 85.468 (+1)	38 **Sr** Strontium 87.62 (+2)	39 **Y** Yttrium 88.906 (+3)	40 **Zr** Zirconium 91.224 (+4)	41 **Nb** Niobium 92.906 (+5)	42 **Mo** Molybdenum 95.95 (+6)	43 **Tc** Technetium 98.907 (+7,+4)	44 **Ru** Ruthenium 101.07 (+4,+3)	45 **Rh** Rhodium 102.906 (+3)	46 **Pd** Palladium 106.42 (+4,+2)	47 **Ag** Silver 107.868 (+1)	48 **Cd** Cadmium 112.414 (+2)	49 **In** Indium 114.818 (+3)	50 **Sn** Tin 118.711 (+4,+2)	51 **Sb** Antimony 121.760 (+5,+3,-3)	52 **Te** Tellurium 127.6 (+6,+4,-2)	53 **I** Iodine 126.904 (+7,+5,+3,+1,-1)	54 **Xe** Xenon 131.294 (+6,+4,+2,0)
6	55 **Cs** Cesium 132.905 (+1)	56 **Ba** Barium 137.328 (+2)	57-71	72 **Hf** Hafnium 178.49 (+4)	73 **Ta** Tantalum 180.948 (+5)	74 **W** Tungsten 183.85 (+6)	75 **Re** Rhenium 186.207 (+7)	76 **Os** Osmium 190.23 (+8)	77 **Ir** Iridium 192.22 (+4,+3)	78 **Pt** Platinum 195.08 (+4,+2)	79 **Au** Gold 196.967 (+3,+1)	80 **Hg** Mercury 200.59 (+2,+1)	81 **Tl** Thallium 204.383 (+3,+1)	82 **Pb** Lead 207.2 (+4,+2)	83 **Bi** Bismuth 208.980 (+3)	84 **Po** Polonium [208.982] (+4,+2)	85 **At** Astatine 209.987 (-1)	86 **Rn** Radon 222.018 (0)
7	87 **Fr** Francium 223.020 (+1)	88 **Ra** Radium 226.025 (+2)	89-103	104 **Rf** Rutherfordium [261] (unknown)	105 **Db** Dubnium [262] (unknown)	106 **Sg** Seaborgium [266] (unknown)	107 **Bh** Bohrium [264] (unknown)	108 **Hs** Hassium [269] (unknown)	109 **Mt** Meitnerium [278] (unknown)	110 **Ds** Darmstadtium [281] (unknown)	111 **Rg** Roentgenium [280] (unknown)	112 **Cn** Copernicium [285] (unknown)	113 **Nh** Nihonium [286] (unknown)	114 **Fl** Flerovium [289] (unknown)	115 **Mc** Moscovium [289] (unknown)	116 **Lv** Livermorium [293] (unknown)	117 **Ts** Tennessine [294] (unknown)	118 **Og** Oganesson [294] (unknown)

Lanthanide Series

57 **La** Lanthanum 138.905 (+3)	58 **Ce** Cerium 140.116 (+4,+3)	59 **Pr** Praseodymium 140.908 (+3)	60 **Nd** Neodymium 144.243 (+3)	61 **Pm** Promethium 144.913 (+3)	62 **Sm** Samarium 150.36 (+3,+2)	63 **Eu** Europium 151.964 (+3,+2)	64 **Gd** Gadolinium 157.25 (+3)	65 **Tb** Terbium 158.925 (+3)	66 **Dy** Dysprosium 162.500 (+3)	67 **Ho** Holmium 164.930 (+3)	68 **Er** Erbium 167.259 (+3)	69 **Tm** Thulium 168.934 (+3)	70 **Yb** Ytterbium 173.055 (+3)	71 **Lu** Lutetium 174.967 (+3)

Actinide Series

89 **Ac** Actinium 227.028 (+3)	90 **Th** Thorium 232.038 (+4)	91 **Pa** Protactinium 231.036 (+5)	92 **U** Uranium 238.029 (+6)	93 **Np** Neptunium 237.048 (+5)	94 **Pu** Plutonium 244.064 (+4)	95 **Am** Americium 243.061 (+3)	96 **Cm** Curium 247.070 (+3)	97 **Bk** Berkelium 247.070 (+3)	98 **Cf** Californium 251.080 (+3)	99 **Es** Einsteinium [254] (+3)	100 **Fm** Fermium 257.095 (+3)	101 **Md** Mendelevium 258.1 (+3)	102 **No** Nobelium 259.101 (+3,+2)	103 **Lr** Lawrencium [262] (+3)

ACS General Chemistry Review provides comprehensive and targeted coverage of general chemistry topics tested on ACS. The content covers foundational principles and theories necessary to understand the material and answer test questions.

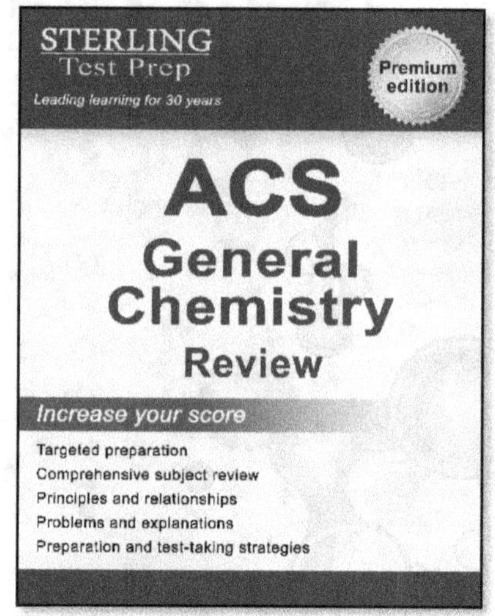

ACS General Chemistry Practice Questions provides high-yield practice questions covering topics tested on ACS General Chemistry exam. Develop the ability to apply your knowledge and quickly choose the correct answer to increase your test score.

Visit our Amazon store

Frank J. Addivinola, Ph.D.

The lead author and chief editor of this study guide is Dr. Frank Addivinola. With his outstanding education, laboratory research, and decades of university science teaching, Dr. Addivinola lent his expertise to develop this book.

Dr. Frank Addivinola conducted original research in developmental biology as a doctoral candidate and pre-IRTA fellow in Molecular and Cell Biology at the National Institutes of Health (NIH). His dissertation advisor was Nobel laureate Marshall W. Nirenberg, Chief of the Biochemical Genetics Laboratory at the National Heart, Lung, and Blood Institute (NHLBI). Before NIH, Dr. Addivinola researched prostate cancer in the Cell Growth and Regulation Laboratory of Dr. Arthur Pardee at the Dana Farber Cancer Institute of Harvard Medical School.

Dr. Addivinola holds an undergraduate degree in biology from Williams College. He completed his Masters at Harvard University, Masters in Biotechnology at Johns Hopkins University, and five other graduate degrees at the University of Maryland University College, Suffolk University, and Northeastern University.

During his extensive career, Dr. Addivinola held faculty positions at colleges and universities, including Harvard University, Johns Hopkins University, University of Maryland and Northeastern University, and taught numerous undergraduate and graduate-level courses in biology, biochemistry, organic chemistry, inorganic chemistry, anatomy and physiology, medical terminology, nutrition, and medical ethics. He received several awards for his research and presentations.

www.ingramcontent.com/pod-product-compliance
Lightning Source LLC
Chambersburg PA
CBHW080734120726
48001CB00009B/2583